Ocean
Secrets of the deep

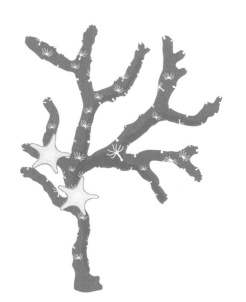

Text by
SABRINA WEISS

Illustrations by
GIULIA DE AMICIS

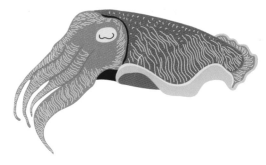

What on Earth Books

Contents

Introduction

More than two thirds of the Earth's surface is covered by seawater. Planet Earth isn't exactly a fitting name for the world we live in. It should really be called Planet Ocean!

As the largest habitat on Earth, the ocean is home to millions of animals and plants. These range from microscopic organisms to the largest animal ever to have lived, the blue whale. The ocean also drives Earth's climate and weather patterns. Seawater absorbs the sun's heat and the ocean currents spread it across the globe.

Now, take a breath.
Take another breath.

Did you know that more than half of the oxygen in our atmosphere comes from the ocean – enough for every second breath we take?

This oxygen gas is produced by tiny marine organisms that live near the water's surface. They use sunlight and carbon dioxide gas to produce energy. This process, photosynthesis, also releases the oxygen we breathe. These marine organisms release more oxygen than all the trees, grasses and other land plants combined. They also form the foundation of nearly all ocean food chains.

Humans need air to breathe, water to drink, food to eat and the right climate to live in. Most of these come from or are brought to us by the ocean.

This means that, no matter how far you are from the shore, the ocean affects your life, and your life affects the ocean, too.

Water is essential to all life on Earth. Every cell in our bodies needs water in order to function and stay healthy. On average, 60 per cent of a human adult is made up of water. This is true for most animals as well.

But the ocean is much more than just a vast expanse of water!

From bright and beautiful coral reefs to undersea forests, from swampy coastal marshes to icy polar seas, the ocean hosts some of the most diverse habitats on our planet. Each provides food and shelter to a huge variety of marine life.

And far below the water's surface, we also find mountains and volcanoes. There is even life at the bottom of the ocean, where strange creatures have learned to live in the pitch-black dark. Their unusual body structures help them survive freezing waters and extreme pressure.

These are truly extreme conditions to get used to.

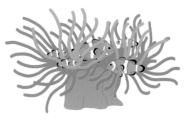

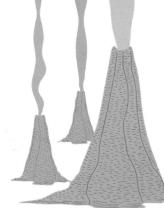

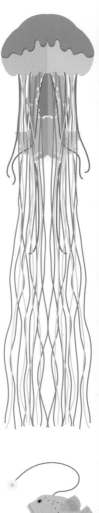

**This book will take you on a journey
to explore a wonderful and mysterious
underwater world, full of unique animals
and plants. You'll learn fascinating facts
and figures as you flick through its pages.**

Inside you will find maps, diagrams and colourful
illustrations that tell a visual story about life
in the ocean.

Most of the ocean still remains a secret. Even so,
new discoveries are being made every day. Humans
are exploring the remotest parts of our blue planet
and learning more and more about the deepest
reaches of the oceans.

**Why is the ocean blue?
Is it reflecting the colour of the sky?**

Not quite. The ocean appears blue because of the
way it filters sunlight.

Sunlight is a mix of red, orange, yellow, green, blue,
indigo and violet light. These colours are the same
as the colours of a rainbow. Each colour is made
up of rays of light with a different wavelength. It's
the value of this wavelength which makes a colour
different from the others. The shortest wavelength is
violet and the longest is red. When sunlight hits the
ocean, the water quickly absorbs colours with long
wavelengths (red, orange and yellow). This leaves the
short-wavelength colours – blue and green – for us
to see.

**The ocean never ceases to amaze.
Let's delve deeper into this remarkable
and fragile environment, and see what
we can discover.**

So, are you ready?

**Take a deep breath and
let's plunge beneath the waves!**

One Planet, One Ocean

You may think that Earth is divided into five oceans – the Pacific, Atlantic, Indian, Southern (Antarctic) and Arctic Ocean. If you look at a map of the world, they are usually labelled this way. The truth is, they are all connected and part of one huge, global ocean.

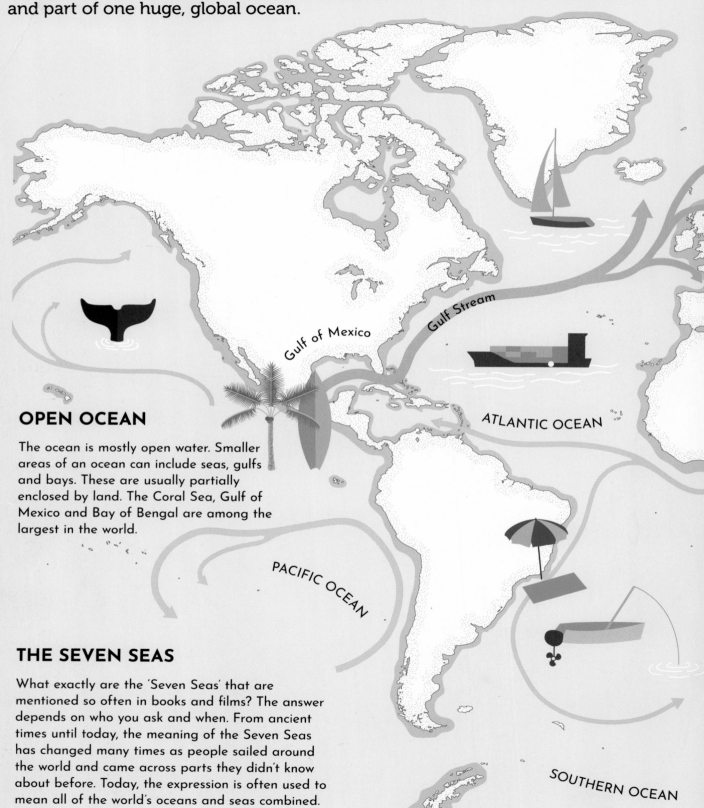

Gulf of Mexico

Gulf Stream

ATLANTIC OCEAN

PACIFIC OCEAN

SOUTHERN OCEAN

OPEN OCEAN

The ocean is mostly open water. Smaller areas of an ocean can include seas, gulfs and bays. These are usually partially enclosed by land. The Coral Sea, Gulf of Mexico and Bay of Bengal are among the largest in the world.

THE SEVEN SEAS

What exactly are the 'Seven Seas' that are mentioned so often in books and films? The answer depends on who you ask and when. From ancient times until today, the meaning of the Seven Seas has changed many times as people sailed around the world and came across parts they didn't know about before. Today, the expression is often used to mean all of the world's oceans and seas combined.

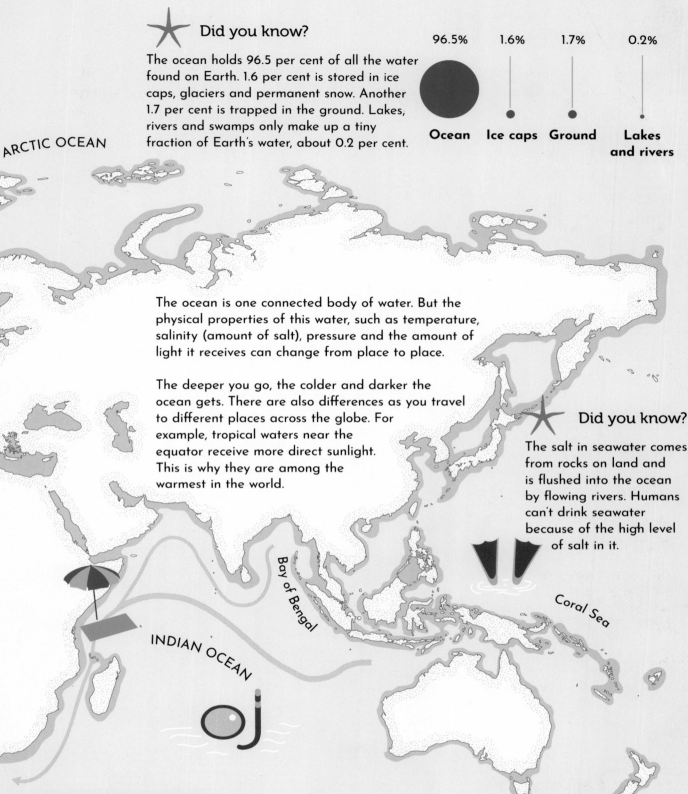

Did you know?

The ocean holds 96.5 per cent of all the water found on Earth. 1.6 per cent is stored in ice caps, glaciers and permanent snow. Another 1.7 per cent is trapped in the ground. Lakes, rivers and swamps only make up a tiny fraction of Earth's water, about 0.2 per cent.

96.5%	1.6%	1.7%	0.2%
Ocean	**Ice caps**	**Ground**	**Lakes and rivers**

ARCTIC OCEAN

The ocean is one connected body of water. But the physical properties of this water, such as temperature, salinity (amount of salt), pressure and the amount of light it receives can change from place to place.

The deeper you go, the colder and darker the ocean gets. There are also differences as you travel to different places across the globe. For example, tropical waters near the equator receive more direct sunlight. This is why they are among the warmest in the world.

Did you know?

The salt in seawater comes from rocks on land and is flushed into the ocean by flowing rivers. Humans can't drink seawater because of the high level of salt in it.

Bay of Bengal

Coral Sea

INDIAN OCEAN

THE OCEAN AND US

The ocean supports life in many ways. Much of the food we eat comes from the ocean. Millions of people rely on fish and shellfish as their main source of protein. A lot of the other food we eat every day contains ingredients that come from the ocean as well. Plus, some substances taken from plants and animals living in the ocean are used to make medicines that can help treat some illnesses.

Legends of the Seas

Throughout the centuries, many myths and legends have grown up around the oceans. Sailors exploring unknown waters, continents and countries returned home with tales full of mysterious creatures they encountered along the way.

THE KRAKEN

This legendary giant squid was said to inhabit the seas between Norway and Greenland. It was feared for attacking large ships and devouring the sailors or fishermen on board. Its arms were so strong that it could pull ships down to the bottom of the ocean.

Greenland

Norway

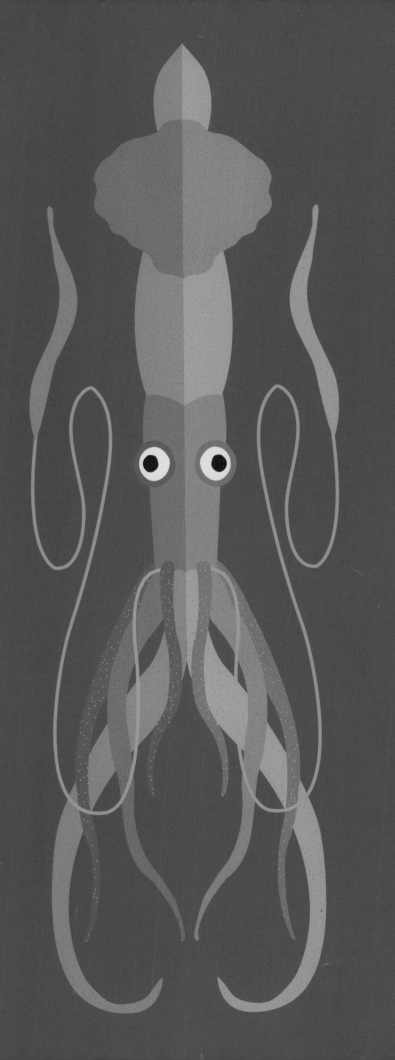

GIANT SQUIDS

There is some truth in this myth – we now know that giant squids exist. But they do not attack ships.

Scientists have been able to study real giant squids that have washed up on shore or have been caught by fishermen around the world. It's thought they can grow 12 m long and roam the ocean at great depths of 300 to 1,000 m.

◄ 0 m

Squids have eight arms, like octopuses. They also have two longer tentacles, which they can shoot out to grab prey.

◄ 200 m

Several different giant squid species have been discovered. But these creatures still remain one of the most elusive large animals. Despite their great size, they are rarely seen alive.

◄ 400 m

◄ 600 m

12 m

◄ 800 m

◄ 1,000 m

MERMAIDS AND MANATEES

Mermaids are legendary half-woman, half-fish creatures. They have been portrayed in ancient stories as well as books and films. It's not clear where and when legends about them originated, but they go back at least to the time of the ancient Greeks. Since then, many sailors and explorers have reported sightings of this incredible creature.

As with many legends, real animals might have inspired stories about mermaids. On his voyage to what is now the Dominican Republic, the explorer Christopher Columbus spotted three mermaids. In fact, they were probably manatees.

Manatees and their close relatives, dugongs, are large, slow-swimming animals that breathe air at the water's surface. Manatees and dugongs are the only vegetarian marine mammals. They love to eat seagrass, algae and freshwater plants.

2.6 m

 Did you know?

Although manatees are sometimes called sea cows, their closest relative on land is the elephant.

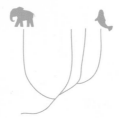

ATLANTIS

The ocean hides many submerged ruins and forgotten shipwrecks. It is a treasure trove of ancient mysteries.

Atlantis was a peaceful island where humans and animals lived in harmony. After a day and a night of earthquakes and floods, the city sank into the depths of the ocean. Or so the story goes.

This tale was described more than 2,300 years ago by the ancient Greek philosopher Plato. The legend of Atlantis has fuelled the imagination of scientists and spawned many theories about the origin of the story. One theory connects it to the volcanic island of Santorini in Greece, one to the Strait of Gibraltar and another to the Black Sea next to Russia.

Scientists have scoured the ocean floor for evidence but so far have had no luck finding anything. There are no records of it anywhere else in the world either.

Some think that Plato didn't even believe the story. They think he invented it to remind people in his wealthy home city of Athens that even the greatest empires can fall.

On the map below are some of the many locations that have been proposed for Atlantis.

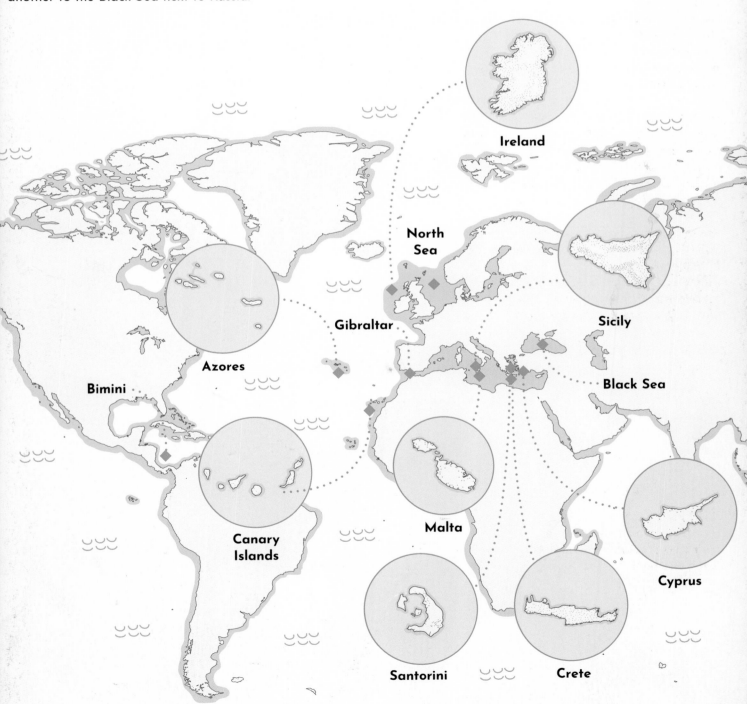

Ireland

North Sea

Azores

Bimini

Gibraltar

Sicily

Black Sea

Canary Islands

Malta

Cyprus

Santorini

Crete

13

Voyage to the Deep

Using new technologies, scientists learn more about the ocean every day. Still, around 90 per cent of the ocean remains unseen by human eyes.

Pressure increases tremendously the deeper underwater we go. So it is difficult for humans to study the oceans at great depths. We actually know more about the surface of the Moon and planets such as Mars than we do about the ocean floor.

Sound travels faster and further in seawater than in air. Scientists use sonar devices to navigate underwater, calculate the size and distance of objects and map the ocean floor.

How deep is the ocean? You'll find out in the next chapters. And you'll also meet the creatures living in it. The ocean offers many different habitats for organisms to thrive. These environments vary in depth, amount of light and nutrients, temperature and pressure.

SPACE VS THE OCEAN

Unfortunately, there still isn't a detailed map of the whole ocean. The entire ocean floor has been mapped to a maximum resolution of 5 km, which means that all features larger than 5 km are visible. Only 10 per cent to 15 per cent of the ocean floor has been mapped at 100 m resolution.

In contrast, the Martian and lunar (Moon's) surfaces have been mapped completely to a resolution of around 100 m.

Moon	100%
Mars	100%
Venus	98%
Ocean	10%

MAPPED SURFACE AT A 100 M RESOLUTION

SUNLIGHT ZONE
0 to 200 m deep

Sunlight reaches this zone and drives the photosynthesis of ocean plants and tiny living things called phytoplankton. But by 100 m below the water's surface, most of the light is gone.

Phytoplankton are single-celled organisms that drift near the surface. They require sunlight to live and grow and are the basis of the marine food chain. Photosynthesis gives off oxygen. Ocean plants and phytoplankton provide more than half of the oxygen humans and other land-dwelling animals breathe.

Marine plants and phytoplankton absorb energy from sunlight and release oxygen into the atmosphere.

-100 m ◀

1. Moray eel
2. Spotted eagle ray
3. Cuttlefish
4. Beadlet anemone
5. Ocean sunfish
6. Common crab
7. Common starfish
8. Parrotfish
9. Sun starfish
10. Red cushion starfish
11. Red coral
12. Fringed sand dollar
13. Common lobster

-200 m ◀

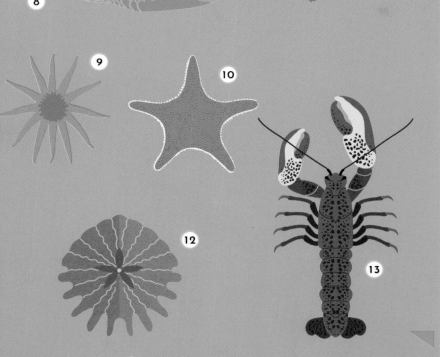

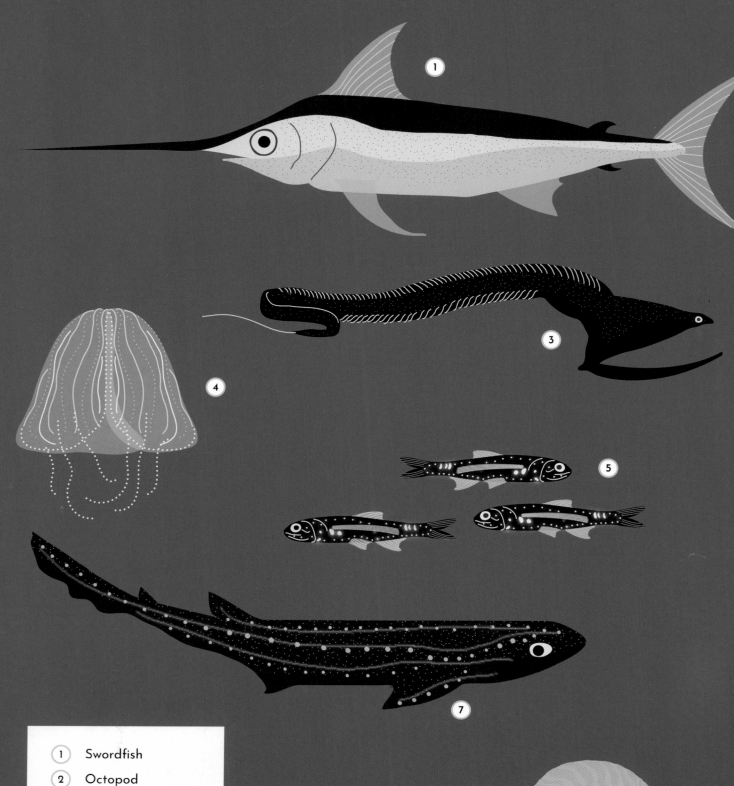
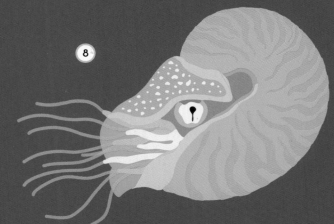

1. Swordfish
2. Octopod
3. Black loosejaw
4. Comb jelly
5. Lanternfish
6. Whipnose anglers
7. Chain catshark
8. Chambered nautilus
9. Glow worms

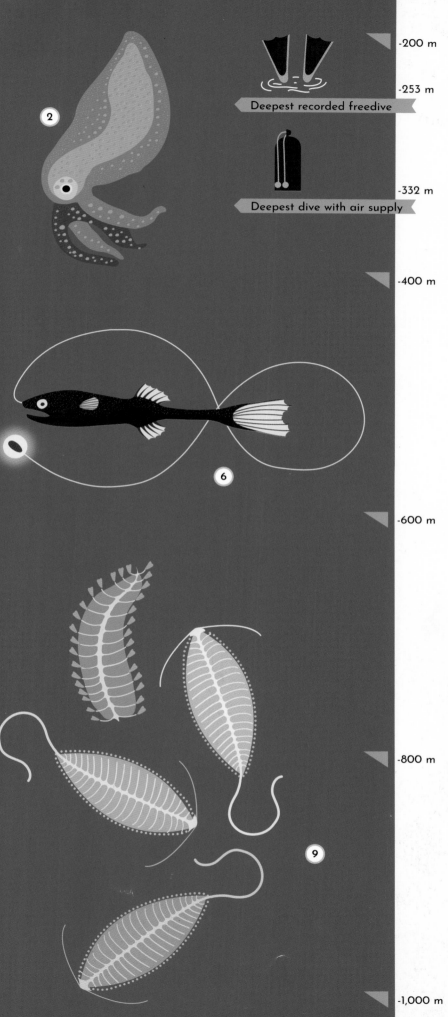

-200 m

-253 m

Deepest recorded freedive

-332 m

Deepest dive with air supply

-400 m

-600 m

-800 m

-1,000 m

TWILIGHT ZONE
200 to 1,000 m deep

Although there is still a little sunlight in the twilight zone during the day, it's not enough to make photosynthesis possible. So plants and algae can't survive at this depth.

Many animals in the twilight zone have huge eyes to capture the little light that remains. Others are 'bioluminescent' – they produce their own light, or host bacteria that give off light for them. Some use this light to blind or distract hungry attackers, to search for food or to communicate.

One bioluminescent animal is the lanternfish. Tiny light-emitting organs called photophores are located on its head, underside and tail. At night, the lanternfish comes close to the surface to eat plankton. But it spends the day in the deep, dark waters.

Chain catsharks are small sharks that hide in nooks and crannies. A special pigment in their skin absorbs blue light and re-emits it as a glowing green. The sharks use this light to stand out and communicate with each other.

★ Did you know?

This part of the ocean is challenging for humans to explore. Professional divers have only ventured to these depths using special equipment.

The deepest recorded freedive (on a single breath) is 253 m, using a torpedo-type sled to descend into the waters off Santorini, Greece. Austrian freediver Herbert Nitsch nearly lost his life during this dive and had to learn to walk and talk again afterwards.

Egyptian diver Ahmed Gabr set the world record depth for scuba diving at 332 m in 2014. Although it only took twelve minutes to reach the deepest point, the total dive time was almost fourteen hours. This is because divers have to make safety stops at different depths along the way.

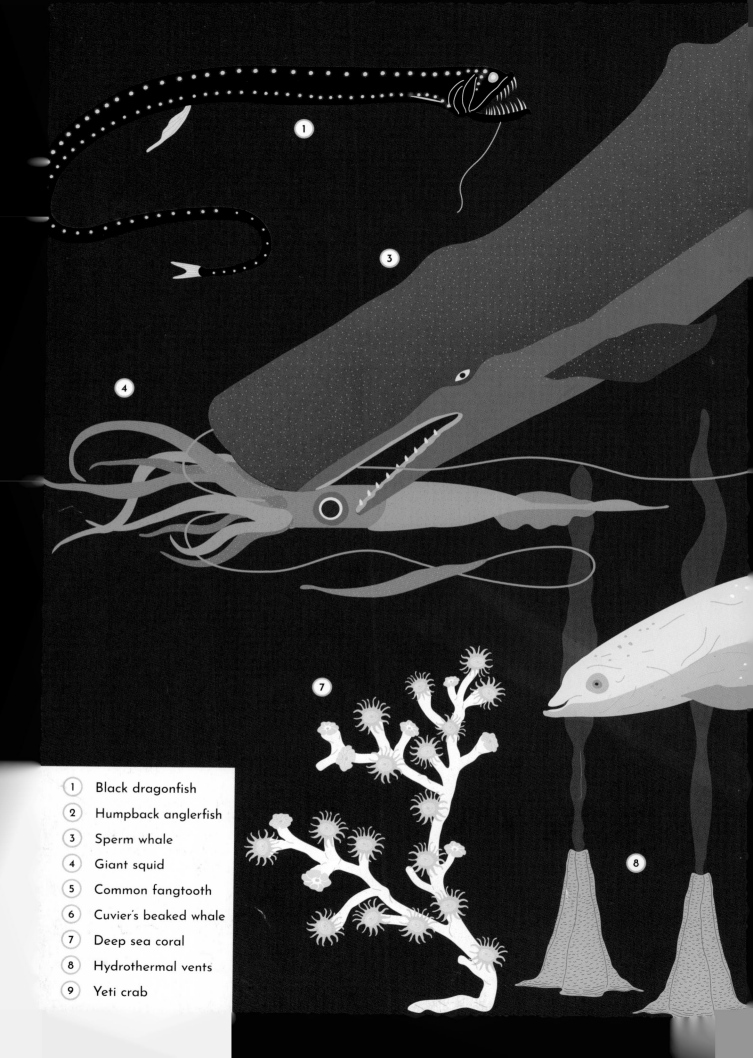

1. Black dragonfish
2. Humpback anglerfish
3. Sperm whale
4. Giant squid
5. Common fangtooth
6. Cuvier's beaked whale
7. Deep sea coral
8. Hydrothermal vents
9. Yeti crab

-1,000 m

MIDNIGHT ZONE
1,000 to 4,000 m deep

Sunlight does not reach the midnight zone. Down here, the ocean is eternally dark. Scientists used to think there could be no life here at all – the environment would be too dark and cold. But some animals have adapted to living in these hostile conditions.

The female humpback anglerfish, for example, relies on glowing bacteria to light her way. Millions of the bacteria live on the end of a fishing rod-like structure attached to her head. Many animals are attracted to light, so this fearsome fish is able to lure its prey close enough to snatch them up in its jaws.

-2,000 m

The male humpback anglerfish is tiny compared to the female. Without its own luminous lure, it relies on its great sense of smell to find its way in the dark.

Like land mammals, marine mammals also need air to breathe. But some are still able to dive this deep in search of food. Sperm whales are known to dive to 2,000 m or more. At these great depths, they hunt giant squid using echolocation.

The Cuvier's beaked whale holds the record among the mammals. It can glide down to nearly 3,000 m and hold its breath underwater for more than two hours.

-3,000 m

Cuvier's beaked whale diving depth

 Did you know?

In 1912, the huge ocean liner RMS *Titanic* collided with an iceberg in the North Atlantic and sank. Today, the shipwreck lies at a depth of around 3,800 m.

-3,800 m

Titanic wreck

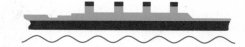

-4,000 m

19

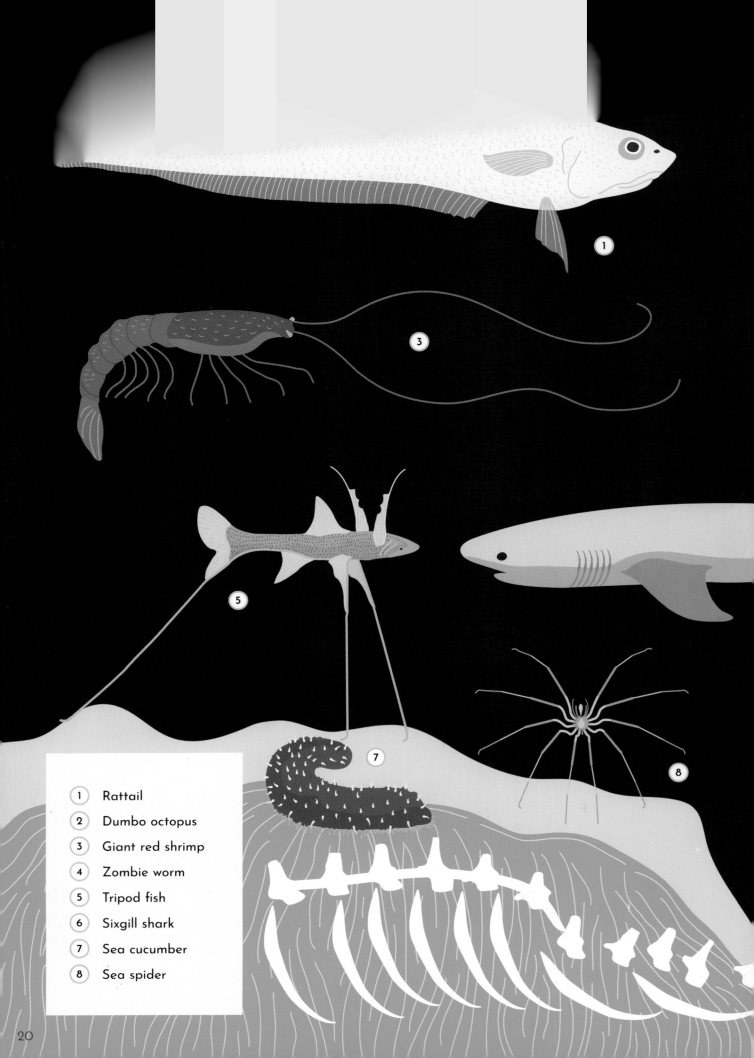

1 Rattail
2 Dumbo octopus
3 Giant red shrimp
4 Zombie worm
5 Tripod fish
6 Sixgill shark
7 Sea cucumber
8 Sea spider

-4,000 m

THE ABYSS
4,000 to 6,000 m deep

The abyssal plains are vast, mostly flat areas of the ocean floor. Strange sea cucumbers, rattails, tripod fish, brittle stars, sea spiders and sea sponges survive down here. They feast on food sinking down from above, from tiny bacteria and plankton to dead plants, fish or even the occasional whale.

When a dead whale comes to rest on the seabed, fish, crabs and other scavengers quickly move in to devour it. They feed on the whale's soft flesh for months until only its skeleton remains. Then, zombie worms and bacteria take over and feast on what's left. The whole meal can last for decades.

The ocean floor is not just a huge patch of sand. Volcanoes and mountains form long chains called mid-ocean ridges too. These rise up from the ocean floor where two of Earth's tectonic plates spread apart and hot magma flows out to fill the gap. One of these chains is the Mid-Atlantic Ridge. It's mostly underwater. It's also the longest mountain range in the world!

-5,000 m

But when two tectonic plates collide underwater, one is forced to bend down and sink beneath the other. This forms a deep trench which can reach down over 10,000 m deep.

 Did you know?

The movement of tectonic plates causes earthquakes. Big earthquakes or underwater volcanic eruptions can cause giant waves to form, called tsunamis. Out in the open ocean, these waves start small. But as they travel towards shallow water they grow and grow. In 1958, a megatsunami hit the coast of Alaska with waves 524 m tall!

Megatsunami
524 m

Empire State Building
381 m

-6,000 m

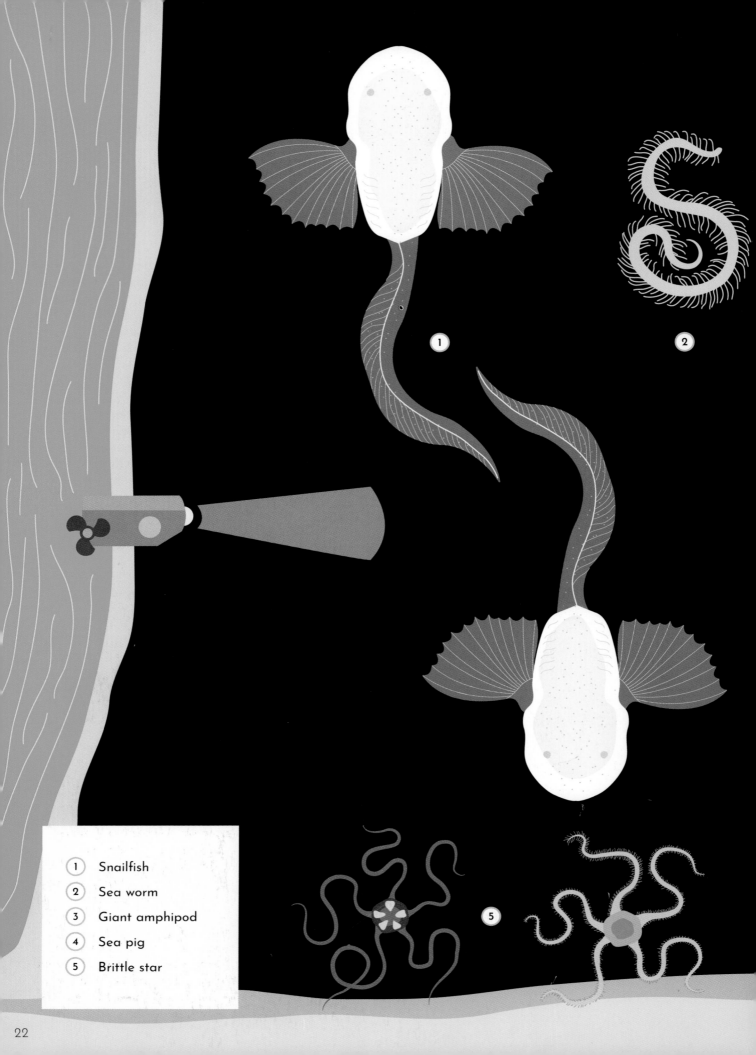

1 Snailfish
2 Sea worm
3 Giant amphipod
4 Sea pig
5 Brittle star

-6,000 m

HADAL ZONE
6,000 to 11,000 m deep

This zone is characterised by deep trenches. The Mariana Trench in the Pacific Ocean is the deepest known point on Earth. One area of the trench - the Challenger Deep - is 10,994 m deep. For comparison Mount Everest stands 8,848 m above sea level.

Mariana Trench
Philippine Sea

-7,000 m

③

⭐ Did you know?

At the bottom of the Challenger Deep, water pressure is a thousand times greater than atmospheric pressure is at sea level.

-8,000 m

◀ 8,848 m

Mount Everest

Challenger Deep

◀ 0

④

-9,000 m

◀ -10,994 m

Only three explorers have ever travelled this deep. The first were Jacques Piccard and Lieutenant Don Walsh in 1960. Despite a cracked window in their submersible, *Trieste*, they stayed on the bottom for 20 minutes. In 2012, film director James Cameron made the trip to the bottom in a submarine called *Deepsea Challenger*. He spent over four hours filming the ocean floor with two 3D cameras. Unmanned vehicles have also been sent down to collect samples and species. Slowly, we are beginning to understand this mysterious location.

-10,000 m

What lurks in the dark this far down? Giant amphipods grow to 34 cm long. Unusual worms and sea cucumbers, such as the sea pig, also live here. Many types of bacteria can be found as well. Snailfish are the deepest living fish ever recorded.

Deepest known point on Earth

-10,994 m

-11,000 m

Underwater Worlds

On sandy beaches or deep in the ocean, marine ecosystems come in many forms. A huge variety of life can be found in these underwater worlds.

ECO-WHAT?

Ecosystems are the living and non-living things in a particular area. Animals, plants and other organisms depend on each other and their natural environment to survive. This includes the water, sunlight, air, rocks and soil around them. In an ecosystem, each organism has a role to play. If a single species is introduced or disappears, it can throw the whole ecosystem out of balance.

INVASIVE SPECIES

Invasive species are organisms that didn't originate in one of the places where they're now found. When they arrive in a new ecosystem they can cause great damage to its environmental balance.

The lionfish usually inhabits the Indo-Pacific Ocean. However, after humans accidentally released six lionfish from an aquarium, it spread into the Atlantic Ocean too. Lionfish eat and reproduce quickly. Plus, they have deadly venomous spines. As a result, native fish numbers are plummeting as they struggle to survive.

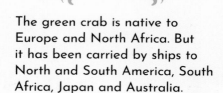

The green crab is native to Europe and North Africa. But it has been carried by ships to North and South America, South Africa, Japan and Australia.

CORAL REEFS

Coral reefs are known for their vibrant colours and the great variety of marine life that lives in them. They are sometimes called the 'rainforests of the ocean'. These ecosystems attract snorkellers and scuba divers from around the world.

Corals are frequently mistaken for rocks or plants. In fact they are made up of hundreds or even thousands of polyps. These are tiny animals related to sea anemones and jellyfish. Coral does not make its own food like plants do. Instead corals have tentacle-like arms for capturing prey from the water.

Corals are living structures made up of hundreds to thousands of polyps.

Corals' bright colours come from tiny algae living inside them.

Some single-celled algae live inside corals. The algae are protected inside the coral and use its waste products for making energy. They combine the waste with sunlight and water to produce sugars for energy. This process is called photosynthesis. It's the same process plants use to make energy.

When the sun shines, the coral gets some of the algae's energy too, so both organisms benefit. This is known as mutualistic symbiosis. But, at night, the coral comes alive in new ways. It stretches out its tentacles to eat any tiny organisms drifting by.

Corals can take many forms. Some are stony, thorny or soft. Others look like feathers, branches or brains! Coral reefs can be small, but the largest can stretch over several hundred kilometres.

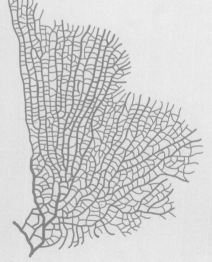

Sea fan coral

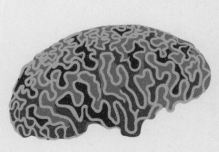

Brain coral

Mushroom coral

THE GREAT BARRIER REEF

This huge system of reefs in northeast Australia is the largest living structure on Earth. It sprawls over 2,300 km and 3,000 individual reefs. The Great Barrier Reef is home to a huge variety of aquatic species. It also provides food for many kinds of birds.

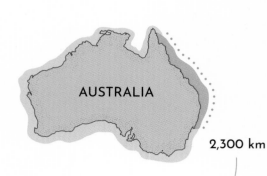

AUSTRALIA

2,300 km

It covers 347,800 sq km, an area larger than the size of Italy.

Number of species supported

The Great Barrier Reef is so massive, it can be seen by astronauts in space.

Did you know?

Coral reefs cover less than 0.2 per cent of the ocean floor. But they support 25 per cent of all known marine life.

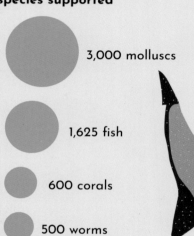

- 3,000 molluscs
- 1,625 fish
- 600 corals
- 500 worms
- 215 birds
- 133 sharks and rays
- 100 jellyfish
- 30 marine mammals

1 — Blacktip reef shark
2 — Moorish idol
3 — Pufferfish
4 — Green turtle
5 — Parrotfish
6 — Butterflyfish

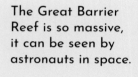
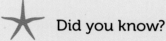

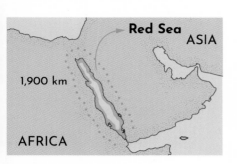

Coral reefs form along the coastlines of the Red Sea between Africa and the Arabian Peninsula.

HUSTLE AND BUSTLE

Coral reefs are like underwater cities where all marine life meets. Many species live permanently on the reef. Some fish, like parrotfish, eat the algae inside the coral. Shrimps and small fish find shelter in the reef's cracks. Sharks and other big fish are frequent visitors to the reef, looking for an easy meal.

Did you know?

Many tropical beaches exist thanks to parrotfish poo. Parrotfish gnaw at the hard shells of coral with their beak-like teeth to get to the algae inside. The broken down coral is indigestible. So the parrotfish get rid of it in the form of sandy poo.

The Great Barrier Reef hosts six of the seven species of sea turtle.

There are almost ninety-five different species of parrotfish in the ocean.

40°C

Coral can survive at temperatures as low as 4°C and as high as 40°C.

4°C

0°C

Shallow-water corals grow in temperatures between 23°C and 29°C. Some corals can tolerate temperatures of as low as 18°C or as high as 40°C for short periods. Cold-water corals do exist as well. They live in deep, dark waters where temperatures range from 4 to 12°C. Like tropical corals, these corals provide a habitat for other species. But they rely completely on food drifting with the currents, rather than on energy from algae.

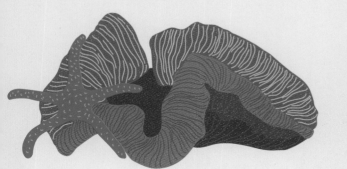

Deltas, Salt Marshes and Mangroves

Coastal ecosystems are found where the land meets the ocean and are home to many species. They also form a natural barrier against waves, protecting the land from flooding and storms.

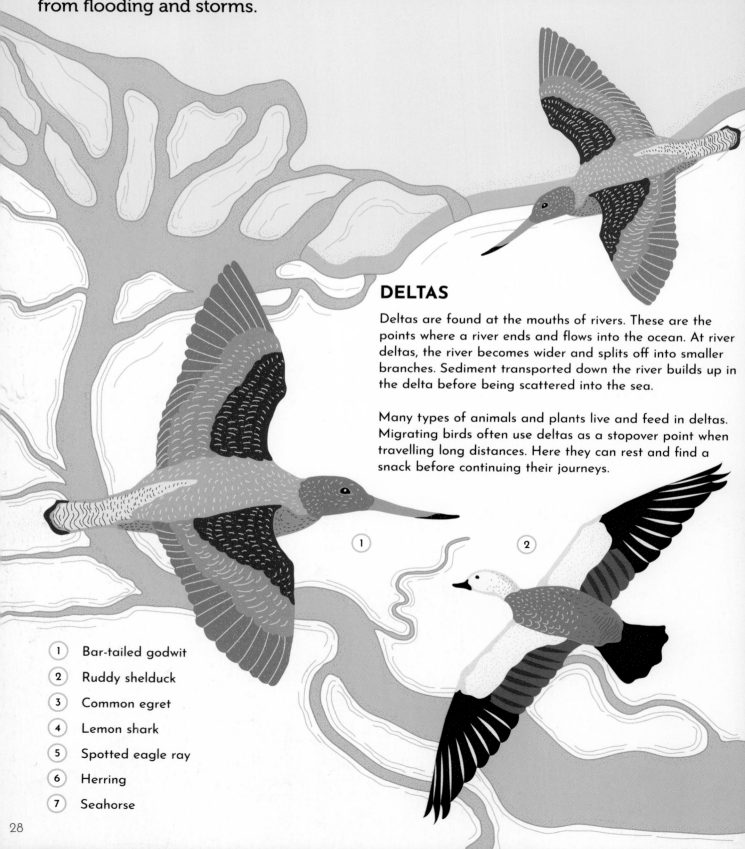

DELTAS

Deltas are found at the mouths of rivers. These are the points where a river ends and flows into the ocean. At river deltas, the river becomes wider and splits off into smaller branches. Sediment transported down the river builds up in the delta before being scattered into the sea.

Many types of animals and plants live and feed in deltas. Migrating birds often use deltas as a stopover point when travelling long distances. Here they can rest and find a snack before continuing their journeys.

1. Bar-tailed godwit
2. Ruddy shelduck
3. Common egret
4. Lemon shark
5. Spotted eagle ray
6. Herring
7. Seahorse

SALT MARSHES

Salt marshes are muddy, sandy areas close to the coast. At high tide, they are flooded by saltwater brought in by the sea. Salt is a challenge for most plants because it dehydrates them. Plants here must be tolerant of salt just to survive.

MANGROVE FORESTS

In tropical regions, mangrove forests form where salt marshes would elsewhere. Mangroves are a family of more than eighty species of trees and shrubs. They are specially adapted to living in salty water.

Like salt marshes, mangrove forests provide a haven for many marine species. The mangrove roots create a maze where young fish, including sharks and rays, can find shelter from predators. Only once they have grown large enough do they venture out into the open ocean.

3

4

5

6

7

29

Kelp Forests

Usually found in shallow waters with plentiful sunlight, kelp creates gorgeous floating forests that provide food and a safe place for many marine species.

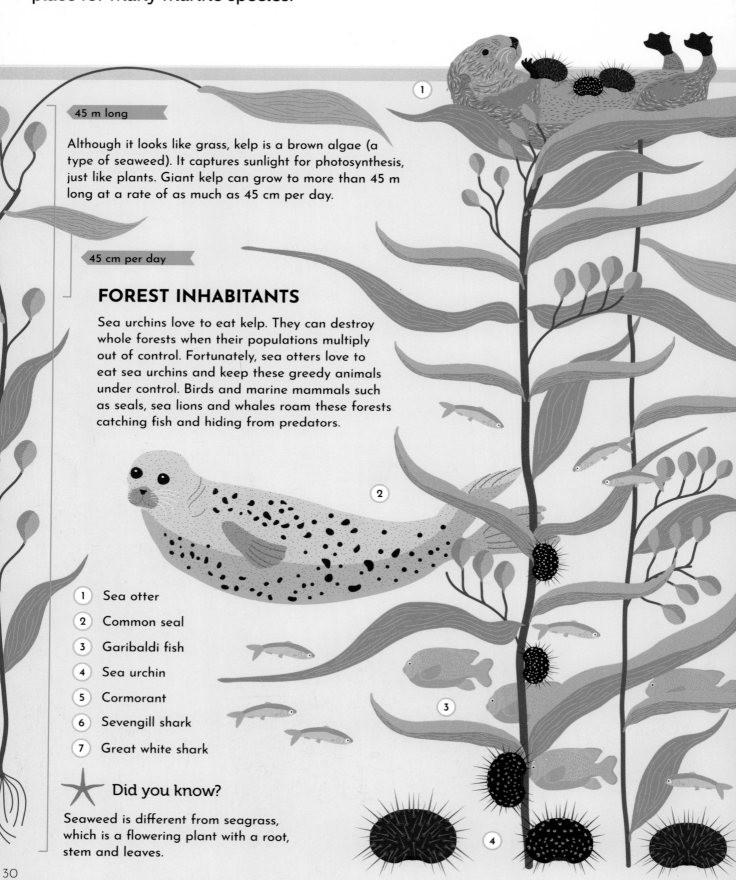

45 m long

Although it looks like grass, kelp is a brown algae (a type of seaweed). It captures sunlight for photosynthesis, just like plants. Giant kelp can grow to more than 45 m long at a rate of as much as 45 cm per day.

45 cm per day

FOREST INHABITANTS

Sea urchins love to eat kelp. They can destroy whole forests when their populations multiply out of control. Fortunately, sea otters love to eat sea urchins and keep these greedy animals under control. Birds and marine mammals such as seals, sea lions and whales roam these forests catching fish and hiding from predators.

1. Sea otter
2. Common seal
3. Garibaldi fish
4. Sea urchin
5. Cormorant
6. Sevengill shark
7. Great white shark

⭐ Did you know?

Seaweed is different from seagrass, which is a flowering plant with a root, stem and leaves.

FEAST FOR BIRDS

Kelp forests are a natural buffet for birds. Cormorants, seagulls, egrets, herons and crows, to name a few, dine on the many fish and invertebrates living inside the forest.

Did you know?

Kelp - particularly its ingredient algin - is valuable for humans. It is used as a thickener or binding agent in ice cream, yoghurt, desserts, toothpaste, shampoo and paint, among many other products. It is also used in medicines.

Some cormorant species have been found to dive as deep as 45 m.

5

KELP AND SHARKS

Some sharks make their home in kelp forests. The broadnose sevengill shark is the most well known. The majority of sharks today have five gill slits on each side. But the broadnose sevengill shark has seven. This seems to be a feature that has remained from some of the earliest sharks. These early sharks lived 400 million years ago, before dinosaurs even appeared on Earth.

400 million years ago

230 million years ago

2 million years ago

6

Sevengill sharks grow up to 3m long. If the desired prey is larger than they are, sevengill sharks will work with others of their kind and hunt as a pack.

These sharks hide between giant towers of seaweed to avoid the feared great white sharks.

7

45 m depth

31

Polar Seas

Polar ecosystems are located at the North and South poles of our planet. Here, in the Arctic and around Antarctica, the waters are bitterly cold. Air temperatures can drop to minus 60°C. Survival in these conditions is tough. But the waters are rich in nutrients and so full of life.

THE ARCTIC

Large areas of the Arctic Ocean are covered year-round with a thick layer of ice. This restricts the amount of sunlight that can reach beneath the surface. The water temperature only just stays above freezing, but some sea dwellers still manage to live here.

The Arctic ice pack will grow in size during the winter months.

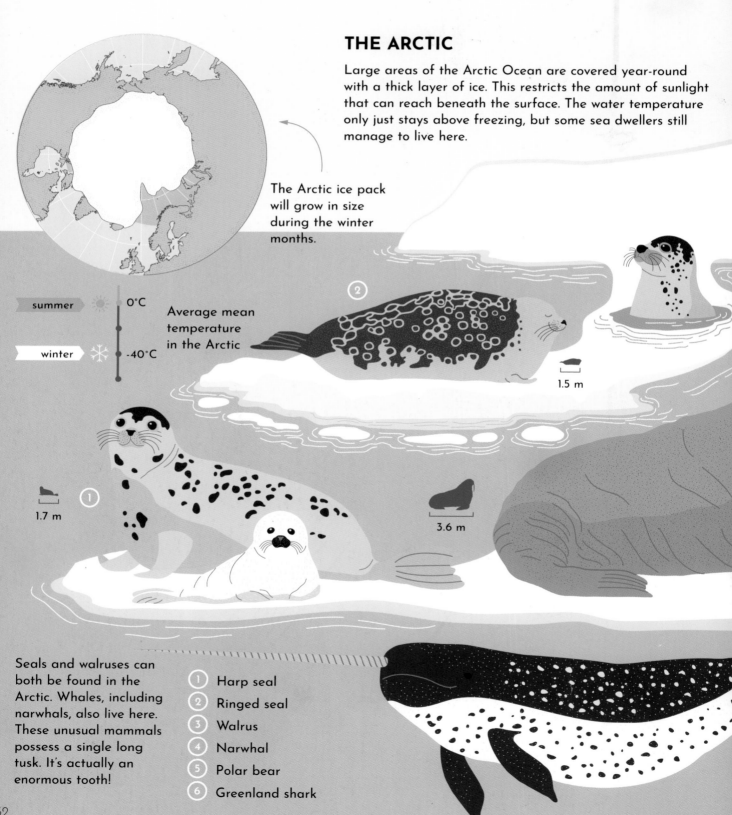

summer — 0°C

winter — -40°C

Average mean temperature in the Arctic

1.5 m

1.7 m

3.6 m

Seals and walruses can both be found in the Arctic. Whales, including narwhals, also live here. These unusual mammals possess a single long tusk. It's actually an enormous tooth!

1. Harp seal
2. Ringed seal
3. Walrus
4. Narwhal
5. Polar bear
6. Greenland shark

POLAR BEARS

Polar bears are the largest carnivores on land, but they also depend on the ocean for survival. They dip in and out of the water to hunt seals, beluga whales and young walruses. Mammals that live in cold waters have developed a special way to stay warm: a thick layer of fat called blubber beneath the skin.

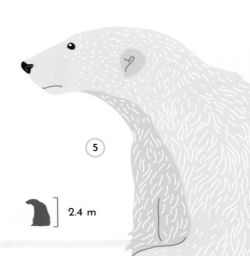

5

2.4 m

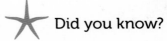

3

skin   blubber muscle

Marine mammals rely on their thick blubber to insulate their bodies in cold water.

4

The narwhal uses its sharp tooth to break the ice.

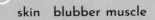

7.5 m

GREENLAND SHARKS

Greenland sharks grow as big as great white sharks and are the largest Arctic fish. These mysterious animals are rarely seen as they live in deep, cold waters. Scientists have suggested that Greenland sharks may be the longest-living vertebrates. They are known to live to at least 272 years old. Some think they could even live to 500!

6

5 m

ANTARCTICA

The North and South Poles are very different places. The Arctic is an icy ocean surrounded by land. But the Antarctic is a huge landmass surrounded by ocean.

Antarctica is situated around the South Pole.

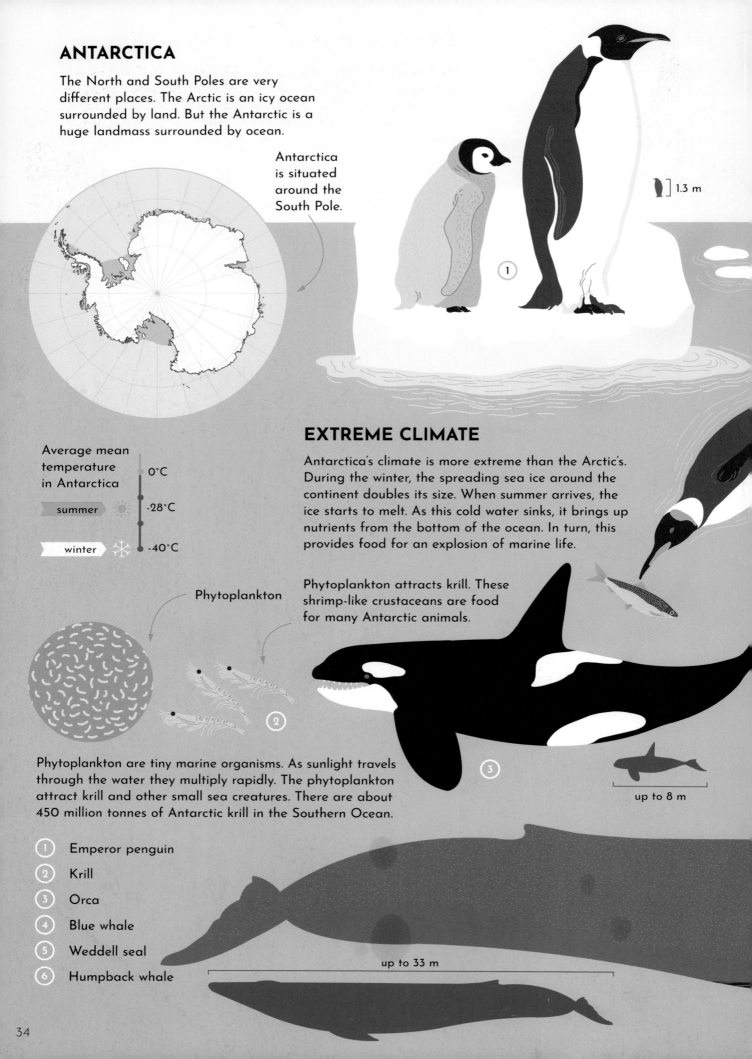

1.3 m

① Emperor penguin

EXTREME CLIMATE

Antarctica's climate is more extreme than the Arctic's. During the winter, the spreading sea ice around the continent doubles its size. When summer arrives, the ice starts to melt. As this cold water sinks, it brings up nutrients from the bottom of the ocean. In turn, this provides food for an explosion of marine life.

Phytoplankton attracts krill. These shrimp-like crustaceans are food for many Antarctic animals.

Average mean temperature in Antarctica

0°C

summer ☀

-28°C

winter ❄

-40°C

Phytoplankton

② Krill

③ Orca

up to 8 m

Phytoplankton are tiny marine organisms. As sunlight travels through the water they multiply rapidly. The phytoplankton attract krill and other small sea creatures. There are about 450 million tonnes of Antarctic krill in the Southern Ocean.

① Emperor penguin
② Krill
③ Orca
④ Blue whale
⑤ Weddell seal
⑥ Humpback whale

up to 33 m

34

A KRILL BUFFET

Most animals living in the waters off Antarctica need krill. If they don't feed on krill directly, then the animals they eat probably do. Seals, penguins, seabirds, fish and squid all depend on krill for survival. Blue and humpback whales travel huge distances just to join the feast.

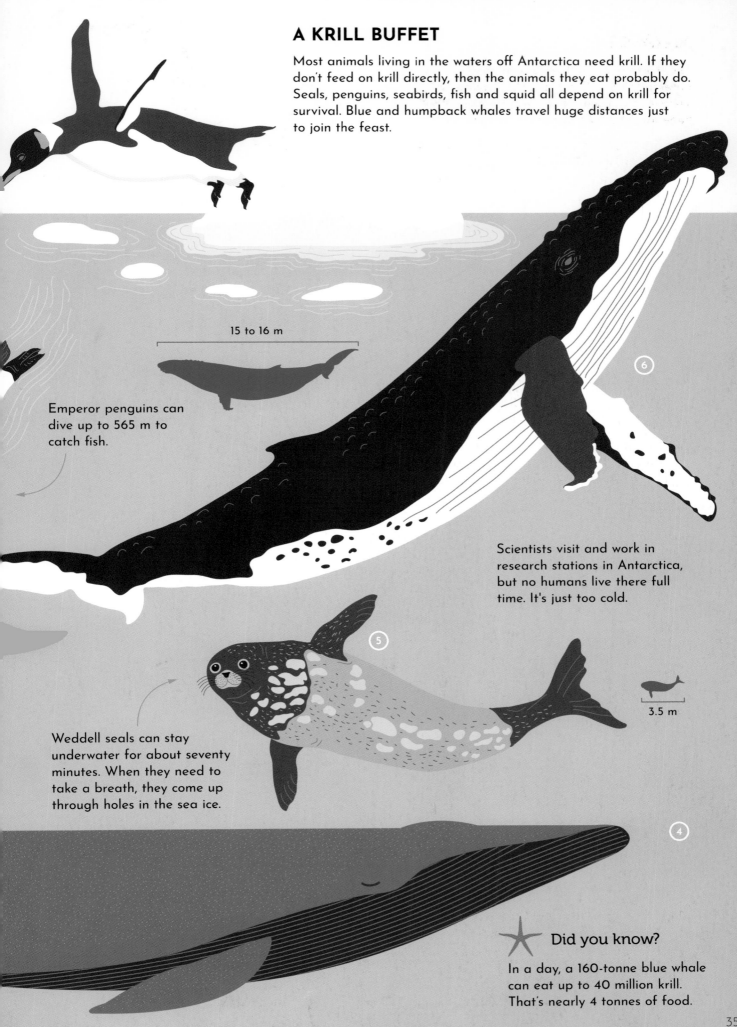

15 to 16 m

Emperor penguins can dive up to 565 m to catch fish.

⑥

Scientists visit and work in research stations in Antarctica, but no humans live there full time. It's just too cold.

⑤

3.5 m

Weddell seals can stay underwater for about seventy minutes. When they need to take a breath, they come up through holes in the sea ice.

④

Did you know?

In a day, a 160-tonne blue whale can eat up to 40 million krill. That's nearly 4 tonnes of food.

Marine Life

Fish come in different shapes, colours and sizes. All fish have some features in common though, including gills, backbones and limbs in the form of fins. Fishes' streamlined bodies minimise drag, making them agile swimmers.

LIFE IN THE OCEAN

Fish have been around much longer than their land-dwelling relatives. Marine life developed 3 billion years earlier than the first life on land. Lizards, frogs, insects, even dinosaurs appeared much more recently. Take a look at the timeline below: the most ancient life is in the ocean.

The majority of species on Earth live in the ocean. From microscopic plankton to fearsome sharks and colossal whales, marine life includes mammals, reptiles, crustaceans, corals, plants, fungi and microbes.

Life in the ocean

Earth forms

4.5 billion years ago

4 billion

First fish

Life on land

First dinosaurs

First species of human

530 million

500 million

230 million

2 millio

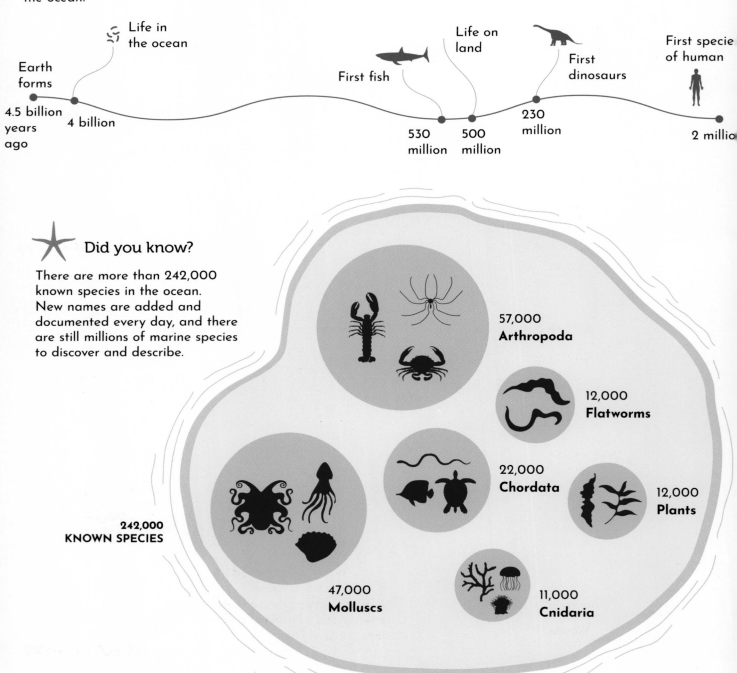

Did you know?

There are more than 242,000 known species in the ocean. New names are added and documented every day, and there are still millions of marine species to discover and describe.

242,000
KNOWN SPECIES

57,000
Arthropoda

12,000
Flatworms

22,000
Chordata

12,000
Plants

47,000
Molluscs

11,000
Cnidaria

WHO EATS WHAT

Organisms in an ecosystem are connected by what they eat and what eats them. Food chains begin with organisms that produce their own food, and end with consumers. So zooplankton feed on phytoplankton, which are producers. In turn, the zooplankton are eaten by small fish. Bigger fish eat the small fish and so on, all the way up to the top of the food chain. When multiple food chains in an ecosystem are connected, this network is called a food web.

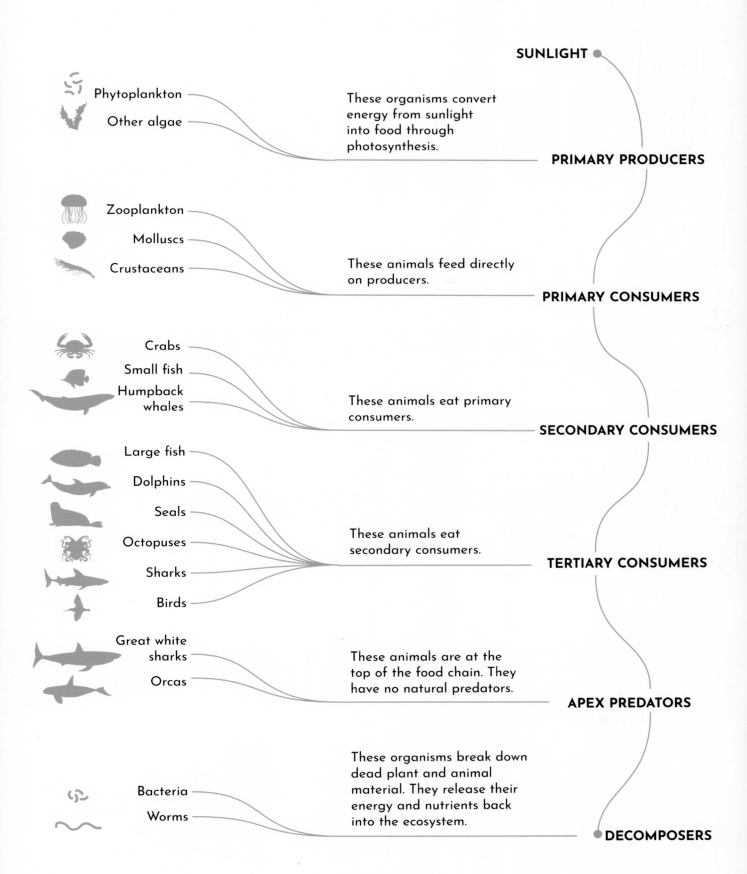

SUNLIGHT

Phytoplankton
Other algae

These organisms convert energy from sunlight into food through photosynthesis.

PRIMARY PRODUCERS

Zooplankton
Molluscs
Crustaceans

These animals feed directly on producers.

PRIMARY CONSUMERS

Crabs
Small fish
Humpback whales

These animals eat primary consumers.

SECONDARY CONSUMERS

Large fish
Dolphins
Seals
Octopuses
Sharks
Birds

These animals eat secondary consumers.

TERTIARY CONSUMERS

Great white sharks
Orcas

These animals are at the top of the food chain. They have no natural predators.

APEX PREDATORS

These organisms break down dead plant and animal material. They release their energy and nutrients back into the ecosystem.

Bacteria
Worms

DECOMPOSERS

Animals as Big as a Bus

From whales and sharks to jellyfish and rays, some sea dwellers make humans look puny. But many of these ocean giants are still poorly understood.

BLUE WHALE

Blue whales are the largest known animals ever to live on Earth. Growing to 33 m long and weighing up to 160 tonnes, these mammals feed almost exclusively on tiny krill - and lots of it! A blue whale can eat 40 million krill per day. These giants are shy and rarely seen by humans.

MEGALODON

Megalodon means 'big tooth' in ancient Greek. Scientists found an 18 cm fossil tooth from this prehistoric shark. From the fossil they estimated its size to be between 14 and 20 m long. This would make it three times as long as the largest great white shark. Megalodon went extinct about 2.6 million years ago but still remains the biggest known shark.

WHALE SHARK

The whale shark is the world's largest living fish. Individuals can grow to around 14 m long and weigh more than 18 tonnes. Despite its name, the whale shark is actually a shark. It has five gills on each side to extract oxygen from the water. See the white spots dotted across its body? Each whale shark has a unique spot pattern. This has allowed researchers to identify different whale sharks.

LION'S MANE JELLYFISH

The bell (main body) of a lion's mane jellyfish is only around 2 m long, but its tentacles can grow to 30 m. Some are even longer than a blue whale. The tentacles are arranged in eight clusters of up to 150 tentacles each. Every tentacle is covered in poisonous stinging cells. The jellyfish use these to catch zooplankton, crustaceans, small fish and other jellyfish.

= 2 m

GIANT SQUID

At 12 m long, giant squids are the largest known invertebrates (animals without backbones). Even so, they mostly remain a mystery. Giant squids' eyes are the largest eyes in the animal kingdom, reaching 25 cm in diameter. That's about the size of a basketball! This is useful in the darkness of the deep ocean. Bigger eyes take in more light and help giant squids detect the shadows of predators, such as the sperm whale.

SPERM WHALE

The sperm whale has an unusual claim to fame – it has the largest brain of any animal! Sperm whales stand out for their distinctive bodies and block-shaped heads. These can measure up to a third of the animal's length. This toothed whale spends a lot of time in deep water, hunting animals like giant squid, octopus and fish.

GREAT WHITE SHARK

One of the most feared ocean predators, the great white shark is at the top of the food chain. A great white's torpedo-shaped body and powerful tail propel it through the water at over 40 km per hour. It can breach (jump) entirely out of the water to attack seals from underneath.

Unusual Friendships

Big groups of fish of the same species that swim together for a purpose, such as hunting, are called schools.

SCHOOLING FISH

A school is a group of fish from the same species that swims in a coordinated pattern. Sticking together in a group can have lots of benefits. Members of the group can find food or a mate more easily and are better protected from predators. When lots of fish move in the same, direction they create a current, which saves each member energy.

Travelling in groups is safer, too. As a large group of fish move quickly to and fro, a predator may become confused and unable to pick out a target. Groups of fish can also be mistaken by predators for a single bigger animal, making them think twice about attacking.

1. Barracuda
2. Giant manta ray
3. Sardine

RAYS

Devil rays can gather in their hundreds and thousands. If they find a plankton-rich patch of ocean, they will come together for a feeding frenzy. Above the waves they can be seen making a splash as they leap out of the water and belly flop.

7 m

GIANT MANTA RAYS

Giant manta rays have a 'finspan' of up to 7 m and are the largest ray species. These curious, social and playful animals are also very intelligent. They have the largest brain relative to their body size of any fish. Giant manta rays like to roam the open ocean in tropical and subtropical waters.

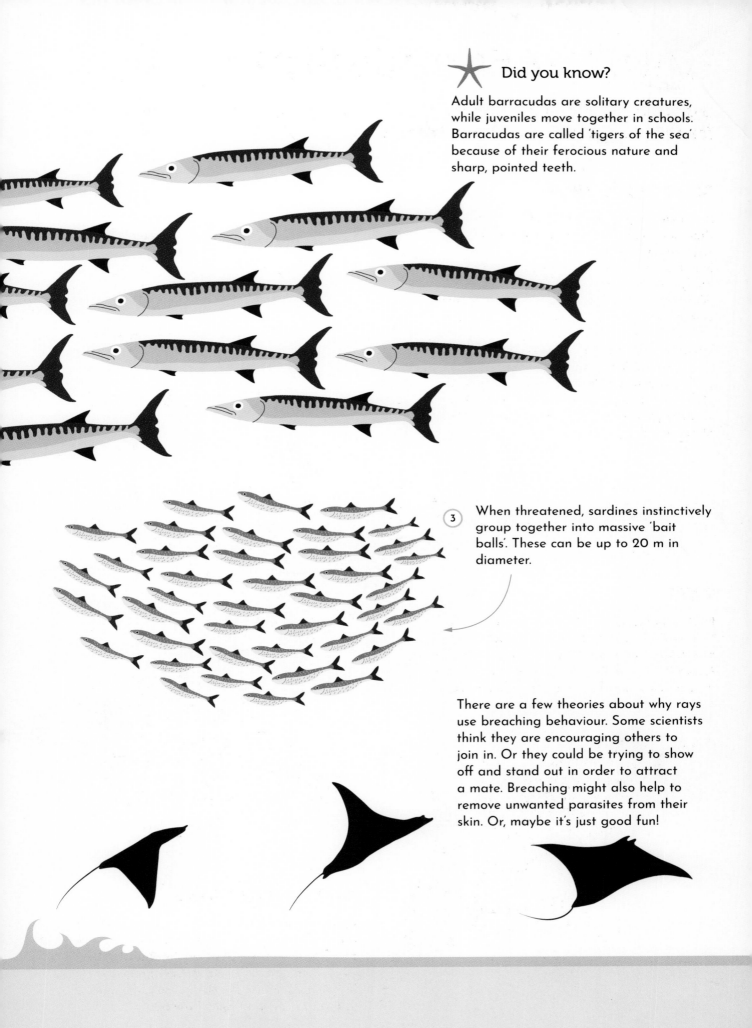

Adult barracudas are solitary creatures, while juveniles move together in schools. Barracudas are called 'tigers of the sea' because of their ferocious nature and sharp, pointed teeth.

3   When threatened, sardines instinctively group together into massive 'bait balls'. These can be up to 20 m in diameter.

There are a few theories about why rays use breaching behaviour. Some scientists think they are encouraging others to join in. Or they could be trying to show off and stand out in order to attract a mate. Breaching might also help to remove unwanted parasites from their skin. Or, maybe it's just good fun!

CLEANING STATIONS

Fish are more social than you might think. They recognise and hang out with each other. They also bond and form long-term relationships, even with other species.

One of these relationships can be seen at 'cleaning stations' across coral reefs. Sea turtles, sharks, rays and other marine creatures turn up here for a free clean.

Cleaners include wrasses, surgeonfish and shrimps. They pick off dead or diseased skin, parasites, algae and slime from their client's body, even from inside its mouth. Then the cleaners eat whatever they've picked up! The client gets a spa treatment, and the cleaners get a meal.

This type of relationship, where both species benefit, is called mutualistic symbiosis.

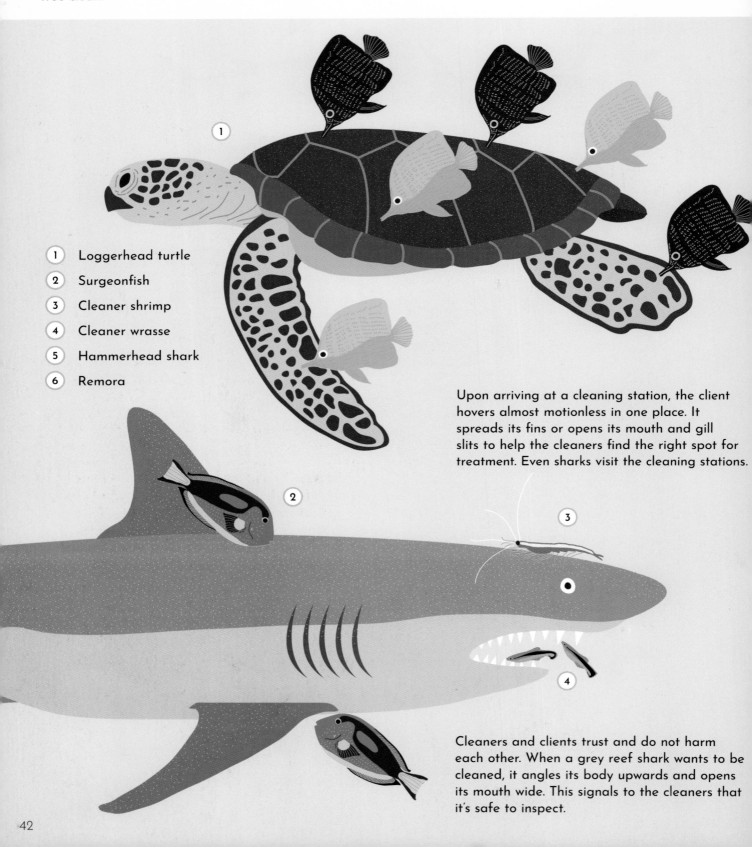

1   Loggerhead turtle

2   Surgeonfish

3   Cleaner shrimp

4   Cleaner wrasse

5   Hammerhead shark

6   Remora

Upon arriving at a cleaning station, the client hovers almost motionless in one place. It spreads its fins or opens its mouth and gill slits to help the cleaners find the right spot for treatment. Even sharks visit the cleaning stations.

Cleaners and clients trust and do not harm each other. When a grey reef shark wants to be cleaned, it angles its body upwards and opens its mouth wide. This signals to the cleaners that it's safe to inspect.

★ **Did you know?**

Some species of cleaners specialise in different parts of their client's body. While some clean the gills, others focus on the mouth, back or fins.

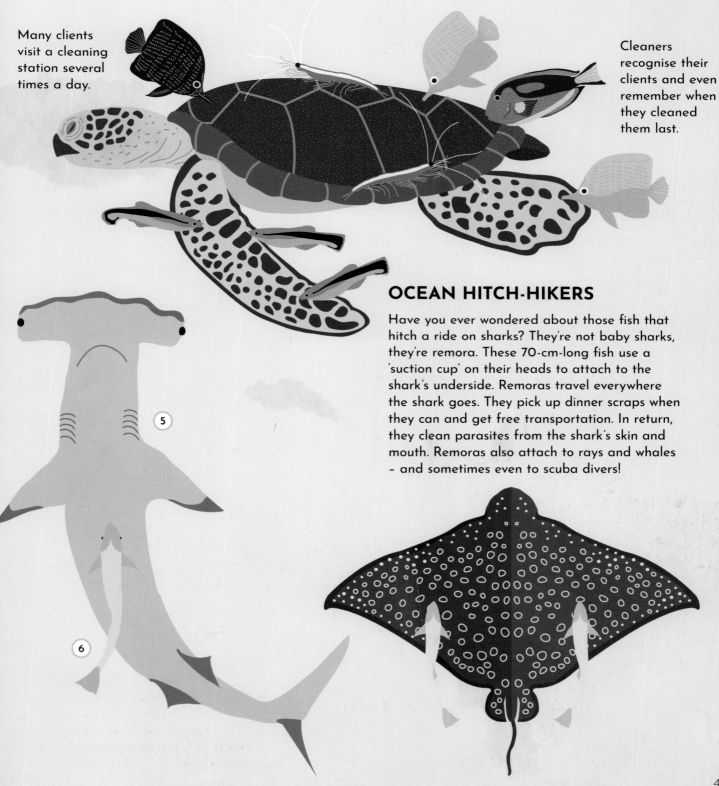

Many clients visit a cleaning station several times a day.

Cleaners recognise their clients and even remember when they cleaned them last.

OCEAN HITCH-HIKERS

Have you ever wondered about those fish that hitch a ride on sharks? They're not baby sharks, they're remora. These 70-cm-long fish use a 'suction cup' on their heads to attach to the shark's underside. Remoras travel everywhere the shark goes. They pick up dinner scraps when they can and get free transportation. In return, they clean parasites from the shark's skin and mouth. Remoras also attach to rays and whales – and sometimes even to scuba divers!

5

6

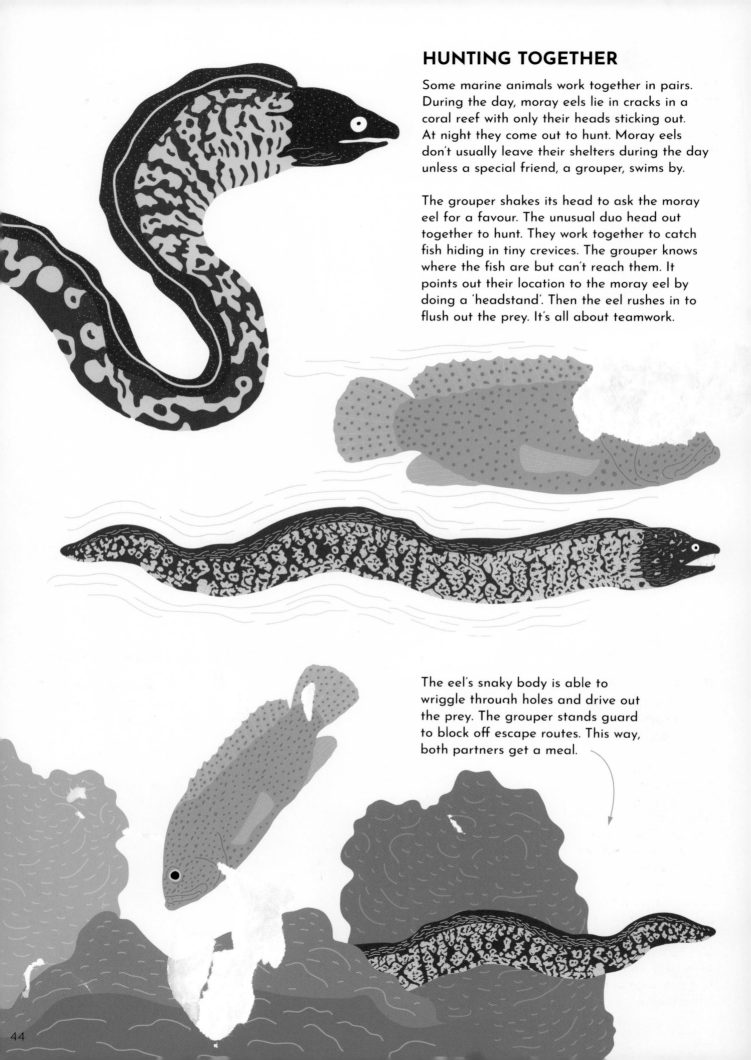

HUNTING TOGETHER

Some marine animals work together in pairs. During the day, moray eels lie in cracks in a coral reef with only their heads sticking out. At night they come out to hunt. Moray eels don't usually leave their shelters during the day unless a special friend, a grouper, swims by.

The grouper shakes its head to ask the moray eel for a favour. The unusual duo head out together to hunt. They work together to catch fish hiding in tiny crevices. The grouper knows where the fish are but can't reach them. It points out their location to the moray eel by doing a 'headstand'. Then the eel rushes in to flush out the prey. It's all about teamwork.

The eel's snaky body is able to wriggle through holes and drive out the prey. The grouper stands guard to block off escape routes. This way, both partners get a meal.

A PECULIAR HOME

Sea anemones have stinging tentacles that are toxic to most animals, but not to clownfish. These little fish are protected by a mucus layer which makes them immune to the anemone's sting.

Clownfish can grow up to 10 cm long. They live six to ten years on average.

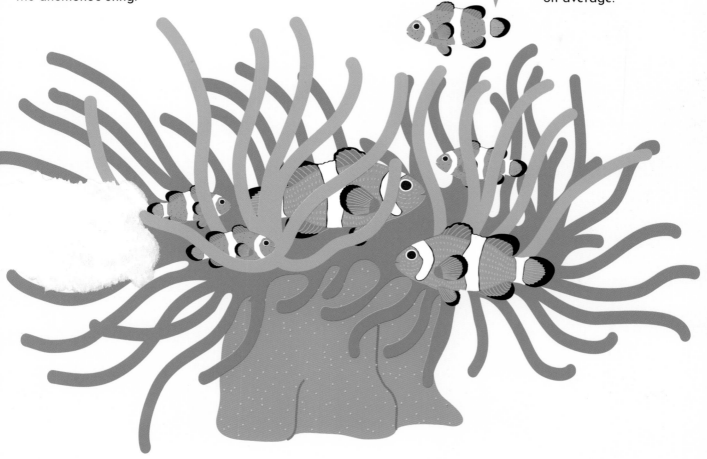

The clownfish live in the anemone. Inside they are protected by its stinging tentacles. In this safe spot, they lay their eggs. Plus, the clownfish gets to eat the anemone's leftover food and parasites. This keeps the anemone healthy and clean too. In return for a safe home, the clownfish protect the anemone from predators. They also help to lure in tasty prey with their bright orange and white bodies.

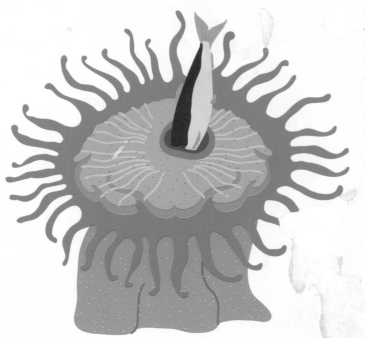

Did you know?

Sea anemones are often called 'flowers of the sea' because of their flowing tentacles and array of colours. But they are not really plants at all. Sea anemones are meat-eating animals that feed on fish, shrimps and jellyfish.

Girl Fish, Boy Fish

Many fish have striking colours and patterns on their bodies. These help them to stand out on the reef and attract a mate.

A BIG EFFORT FOR A TINY FISH

Some species put in a bit of extra effort to get noticed. 12-cm-long male pufferfish found off the coast of Japan definitely like to put on a display. They swim along the ocean floor, flapping their fins to make furrows in the sand. These furrows form a circular pattern about 2 m in diameter. Then the pufferfish pick up shells to decorate the ridges of their construction. They take such care that it can take several days to complete a single one of these masterpieces. The pufferfish make these patterns to charm passing females. If a female is interested, she lays her eggs in the furrows to keep them safe from the currents, before moving on. The male sticks around until the eggs hatch.

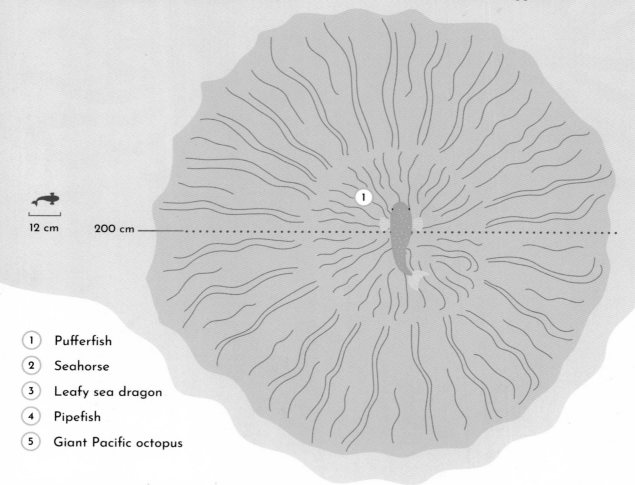

12 cm   200 cm

1　Pufferfish

2　Seahorse

3　Leafy sea dragon

4　Pipefish

5　Giant Pacific octopus

CHANGING SEX

Differentiating between female and male fish is sometimes not so easy. Clownfish, moray eels, groupers and some other fish species are able to change their sex. They can switch their size, colour and reproductive organs during their lifetime in order to ensure they get a chance to reproduce.

Clownfish usually live in groups of one breeding couple and several smaller individuals. The largest fish is female and all others are males or juveniles.

If the female dies, the alpha male changes sex and takes her place. It forms a new couple with the next largest male in the group.

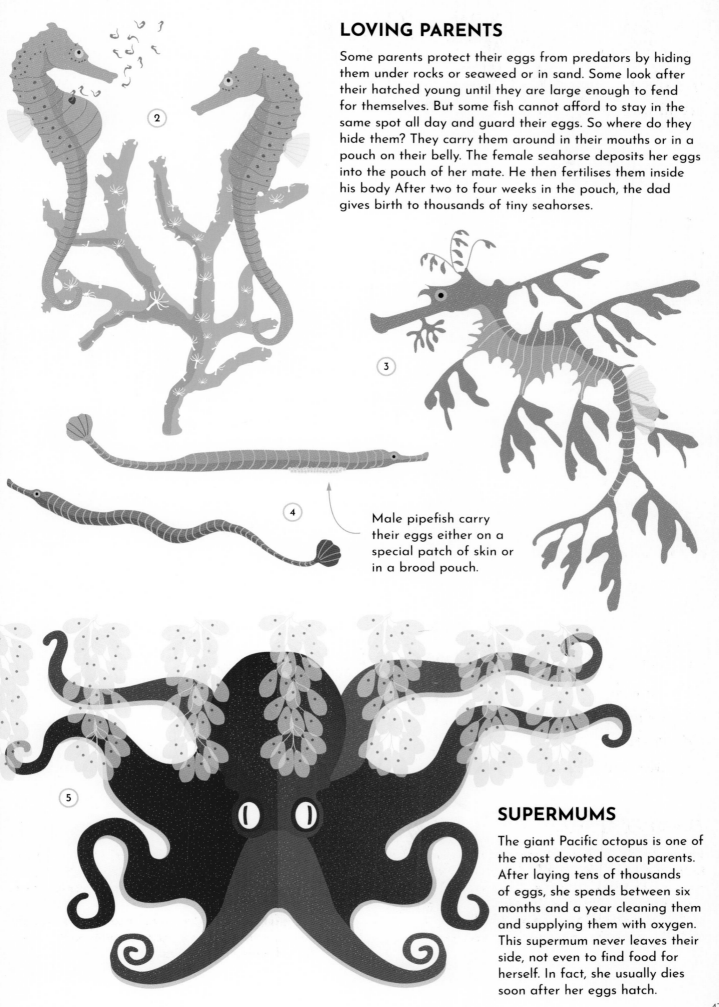

LOVING PARENTS

Some parents protect their eggs from predators by hiding them under rocks or seaweed or in sand. Some look after their hatched young until they are large enough to fend for themselves. But some fish cannot afford to stay in the same spot all day and guard their eggs. So where do they hide them? They carry them around in their mouths or in a pouch on their belly. The female seahorse deposits her eggs into the pouch of her mate. He then fertilises them inside his body After two to four weeks in the pouch, the dad gives birth to thousands of tiny seahorses.

Male pipefish carry their eggs either on a special patch of skin or in a brood pouch.

SUPERMUMS

The giant Pacific octopus is one of the most devoted ocean parents. After laying tens of thousands of eggs, she spends between six months and a year cleaning them and supplying them with oxygen. This supermum never leaves their side, not even to find food for herself. In fact, she usually dies soon after her eggs hatch.

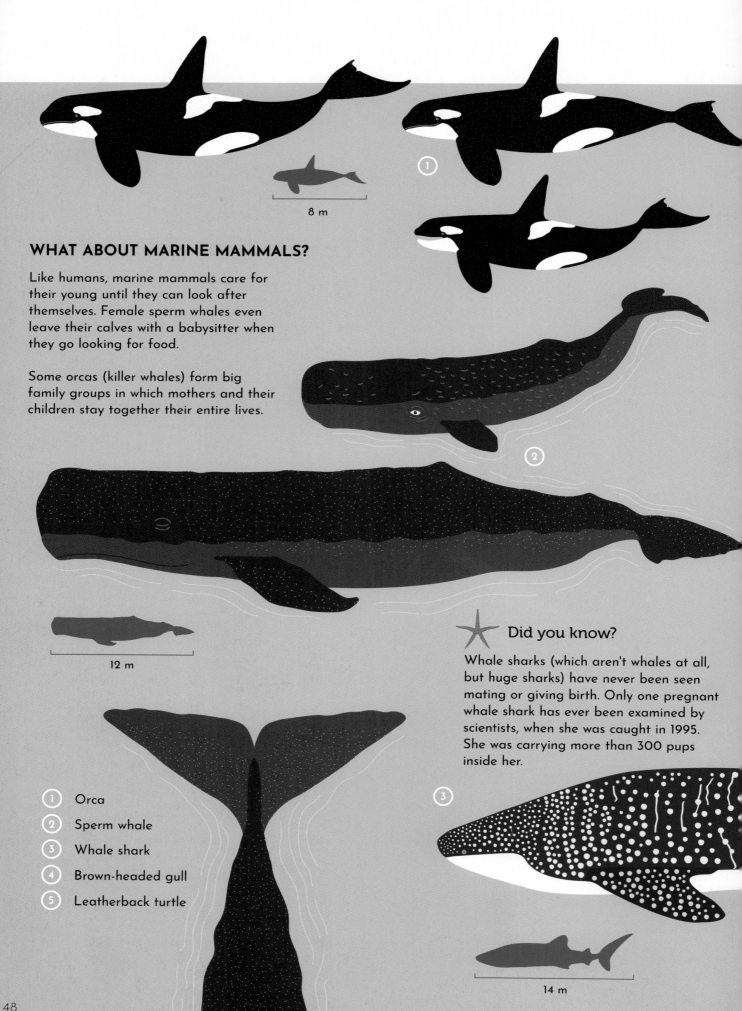

WHAT ABOUT MARINE MAMMALS?

Like humans, marine mammals care for their young until they can look after themselves. Female sperm whales even leave their calves with a babysitter when they go looking for food.

Some orcas (killer whales) form big family groups in which mothers and their children stay together their entire lives.

8 m

12 m

⭐ Did you know?

Whale sharks (which aren't whales at all, but huge sharks) have never been seen mating or giving birth. Only one pregnant whale shark has ever been examined by scientists, when she was caught in 1995. She was carrying more than 300 pups inside her.

1. Orca
2. Sperm whale
3. Whale shark
4. Brown-headed gull
5. Leatherback turtle

14 m

④

1 m

⑤

2.2 m

⭐ Did you know?

When they're ready to lay their eggs, leatherback sea turtles migrate to the same beach where they were born. They use Earth's magnetic field, and probably their keen sense of smell, to guide themselves back to the right spot. Once they arrive, they dig nests in the sand and lay their own eggs inside.

Unlike other species of sea turtle, the leatherback turtle doesn't have a bony shell. Its carapace is actually made of a leather-like skin.

Newborn whale sharks are estimated to be around 60 cm long.

Making Sounds

Marine animals swimming near the surface can see pretty well. But for those in deep water where the light dwindles to little or none, sound becomes much more important. Sound travels through water almost five times faster than it does in air and allows animals to talk to each other over great distances.

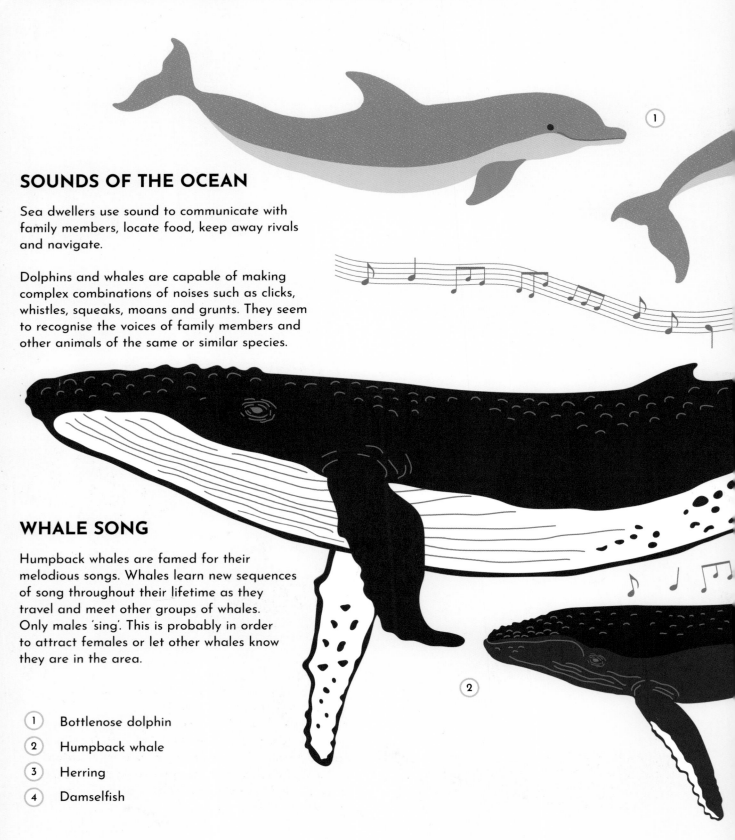

SOUNDS OF THE OCEAN

Sea dwellers use sound to communicate with family members, locate food, keep away rivals and navigate.

Dolphins and whales are capable of making complex combinations of noises such as clicks, whistles, squeaks, moans and grunts. They seem to recognise the voices of family members and other animals of the same or similar species.

WHALE SONG

Humpback whales are famed for their melodious songs. Whales learn new sequences of song throughout their lifetime as they travel and meet other groups of whales. Only males 'sing'. This is probably in order to attract females or let other whales know they are in the area.

① Bottlenose dolphin

② Humpback whale

③ Herring

④ Damselfish

Did you know?

When hunting, dolphins use echolocation to locate food, just as bats do on land. The dolphin makes a sound and then listens for echoes that bounce off prey. The length of time it takes for the echo to return tells the dolphin how far away the prey is.

sound is released

sound bounces off prey and comes back

Humans used to think of the ocean as silent, but it is actually quite noisy. Some reef fish, such as the damselfish, make pulse sounds like pops and chirps to defend their territory. Herrings fart to produce bubbles and create a high-pitched sound. Other fish use teeth grinding, muscle flexing or swim bladder vibrating to deter attackers or attract a mate.

Masters of Disguise

To avoid hungry predators, marine life has developed some very impressive defence mechanisms. Sponges and some corals have spikes and stinging tentacles. Sea stars and urchins use sharp spines. But some fish have to be a bit more creative.

CAMOUFLAGE

Camouflage is when an organism matches its form or colour to its surroundings in order to blend in. Flatfish, for example, can lighten or darken their bodies to match the colour of the sea floor where they are resting.

Other species look like seaweed, plants or corals to avoid being spotted by predators. Ghost pipefish can be hardly visible among algae. Young batfish look like fallen leaves as they hang near the water's surface.

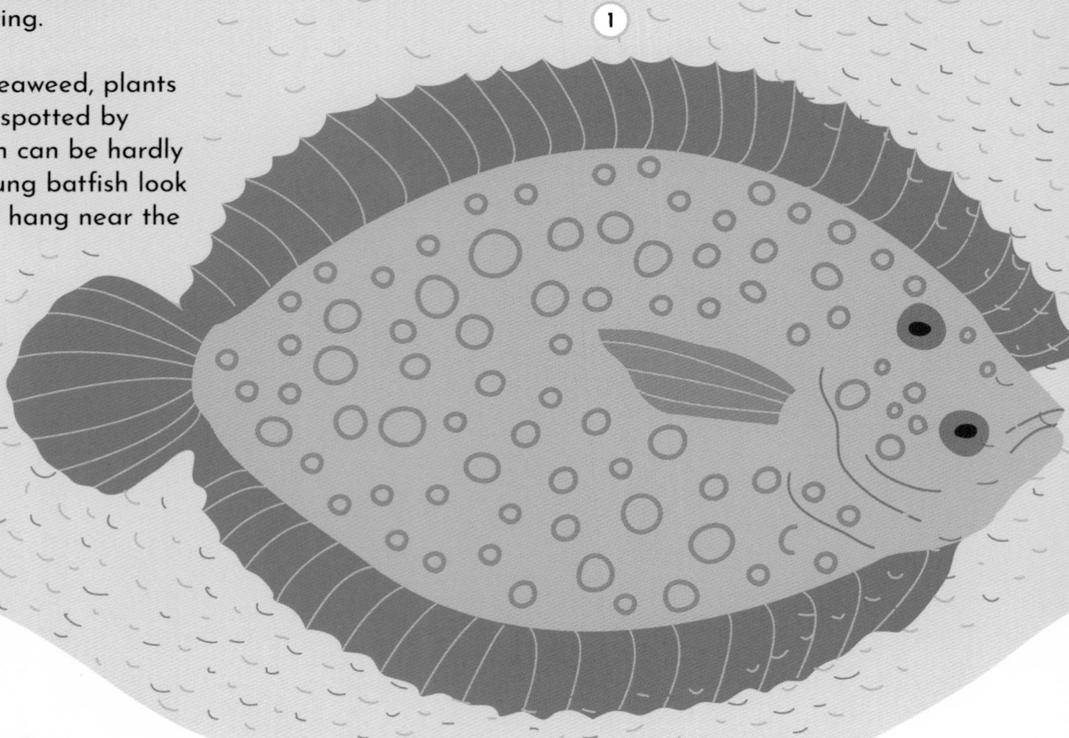

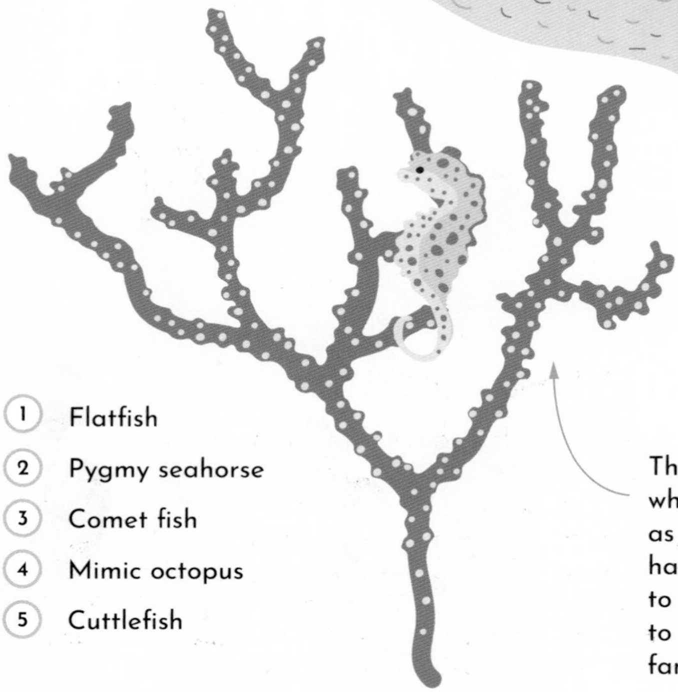

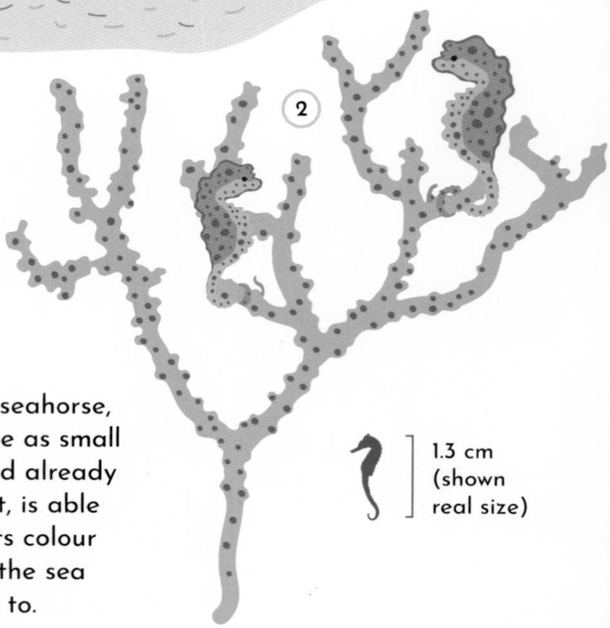

1. Flatfish
2. Pygmy seahorse
3. Comet fish
4. Mimic octopus
5. Cuttlefish

The pygmy seahorse, which can be as small as 1.3 cm and already hard to spot, is able to change its colour to look like the sea fan it clings to.

1.3 cm (shown real size)

MIMICS

Some organisms try a different tactic. They copy the appearance of inedible or dangerous organisms to trick predators into leaving them alone. Some fish mimic the bright colours or swimming style of other marine animals that taste bad, are venomous or have sharp spines.

Some species mimic cleaner fish for safety. Since they perform such a useful function on the reef, most other animals usually leave them alone.

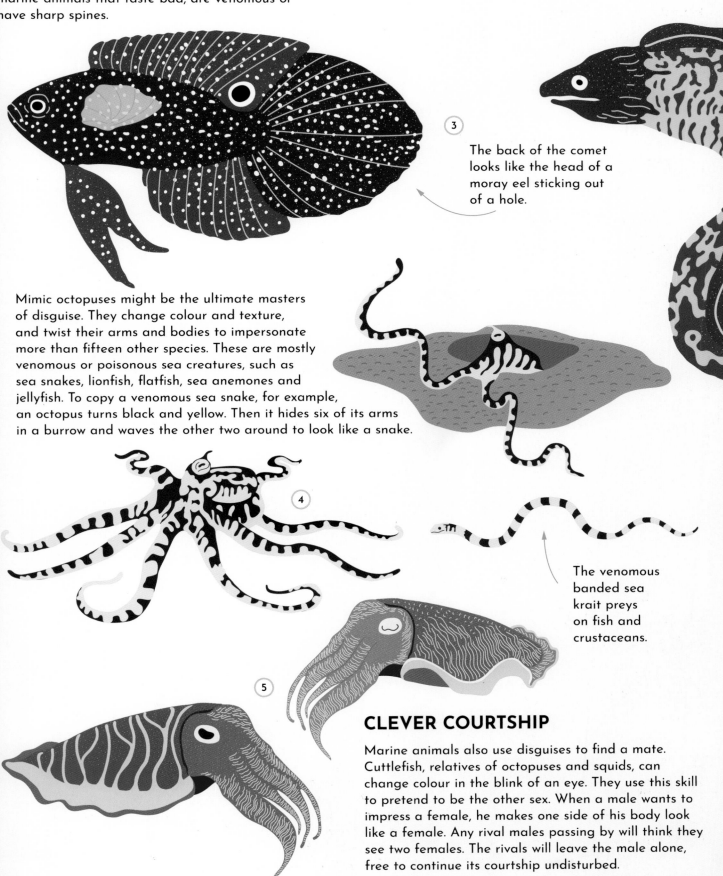

3

The back of the comet looks like the head of a moray eel sticking out of a hole.

Mimic octopuses might be the ultimate masters of disguise. They change colour and texture, and twist their arms and bodies to impersonate more than fifteen other species. These are mostly venomous or poisonous sea creatures, such as sea snakes, lionfish, flatfish, sea anemones and jellyfish. To copy a venomous sea snake, for example, an octopus turns black and yellow. Then it hides six of its arms in a burrow and waves the other two around to look like a snake.

4

5

The venomous banded sea krait preys on fish and crustaceans.

CLEVER COURTSHIP

Marine animals also use disguises to find a mate. Cuttlefish, relatives of octopuses and squids, can change colour in the blink of an eye. They use this skill to pretend to be the other sex. When a male wants to impress a female, he makes one side of his body look like a female. Any rival males passing by will think they see two females. The rivals will leave the male alone, free to continue its courtship undisturbed.

Ocean Wanderers

Most of marine life can be found in nutrient-rich, shallow waters. In these areas, food is abundant. In order to reach these waters, many animals make incredible journeys.

Marine animals will cross the ocean to find a meal and a safe place to mate, spawn, breed and nest.

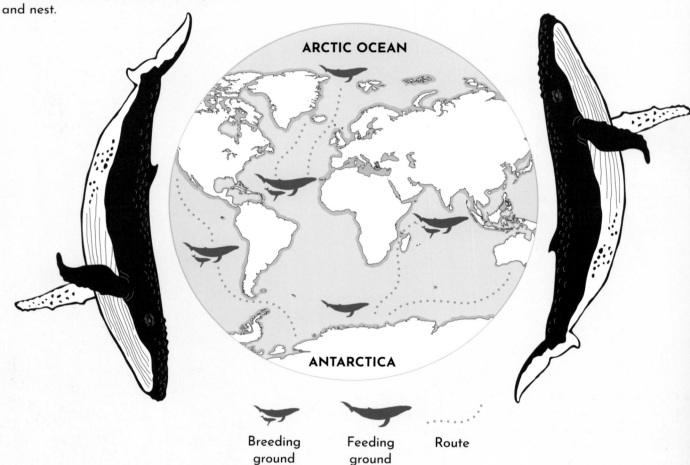

ARCTIC OCEAN

ANTARCTICA

Breeding ground

Feeding ground

Route

A WHALE JOURNEY

Krill-loving humpback whales feed and store up energy in polar regions. This is preparation for an astonishing journey. They migrate around 5,000 km to tropical waters where they mate and give birth. In these waters their young calves are safe from orcas and other predators.

For around six months the adult humpbacks live off the fat reserves stored up during the feeding period. They raise their newborn calves until they are strong enough to make the long trip back to the feeding grounds.

76 cm

UPRIVER MIGRATION

Pacific salmon make an epic round-trip migration. They hatch from their eggs in rivers and streams far from the ocean. After a few weeks to as much as a year, they travel downriver to the ocean, where they spend most of their adult lives.

When it is time to reproduce, the salmon travel hundreds of kilometres upriver to the place where they were born. Most of the exhausted salmon die before their eggs hatch, and the baby salmon start the whole cycle again.

2.3 m

②

③

2 m

During salmon spawning season, some land animals start their own migration. In Alaska, grizzly bears and bald eagles travel to salmon spawning sites. The huge numbers of fish provide a feast for these big predators. The bears eat as much as they can in preparation for their winter hibernation.

① Salmon

② Bald eagle

③ Grizzly bear

 Did you know?

Many marine animals can navigate huge distances in the vast expanse of the ocean. Some use the sun and stars to follow directions. Others orient themselves by underwater mountains or currents. Salmon use Earth's magnetic field to find the rivers where they were born.

Ocean in Peril

For a long time, humans have treated the ocean like a never-ending resource. We have fished it as if it could feed us forever. We have dumped waste in it as if it would never start to fill up. It's now clear that humans are having an impact on nearly all parts of the ocean.

FISH ON A PLATE

Fish and shellfish are a good source of protein. Billions of people around the world depend on them for survival. Once the ocean was full of enough food for everyone who fished it. Today this is no longer the case. Many regions are already overfished. Fish and shellfish are caught in such massive quantities that they don't have time to replenish themselves.

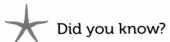 Did you know?

In some cases, overfishing is so serious that species have been classed as in danger of extinction. Bluefin tuna populations are estimated to be down 96 per cent on historic levels. These large fish average 2 m in length and 200 kg in weight. Tuna is a favourite food for many, particularly sushi lovers.

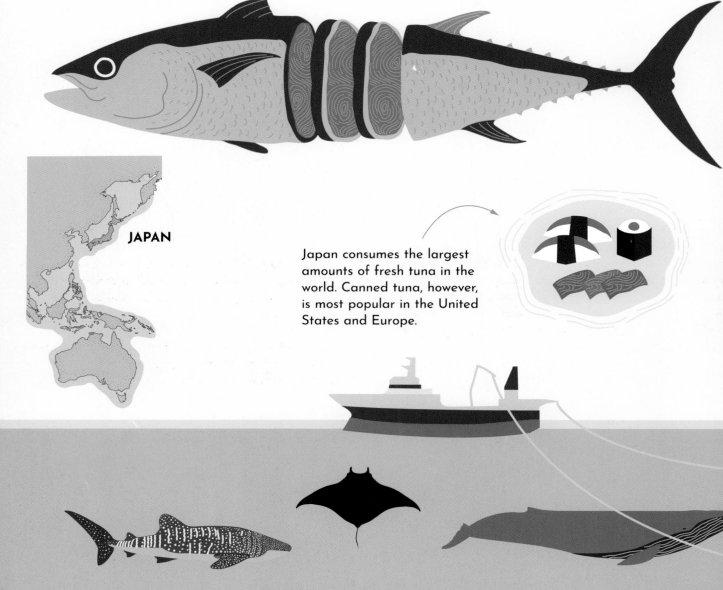

JAPAN

Japan consumes the largest amounts of fresh tuna in the world. Canned tuna, however, is most popular in the United States and Europe.

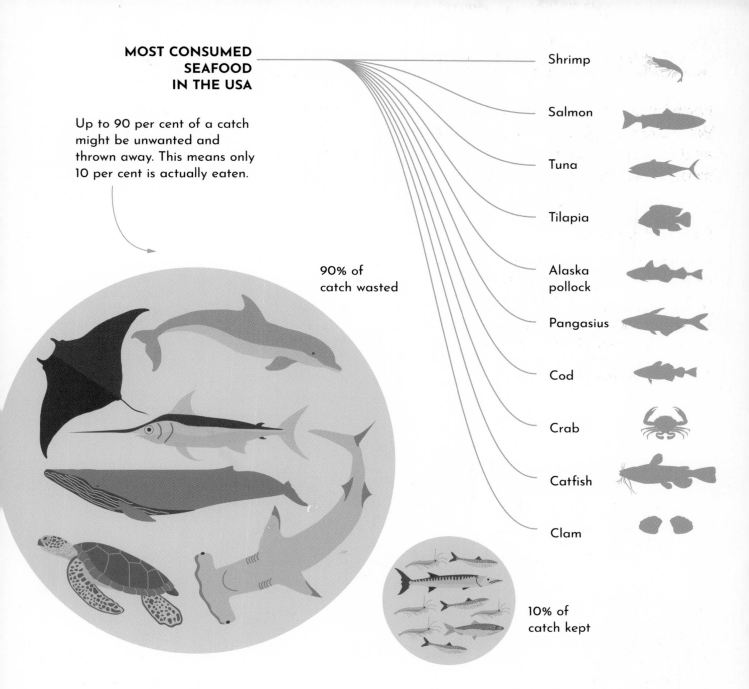

MOST CONSUMED SEAFOOD IN THE USA

Up to 90 per cent of a catch might be unwanted and thrown away. This means only 10 per cent is actually eaten.

90% of catch wasted

10% of catch kept

Shrimp

Salmon

Tuna

Tilapia

Alaska pollock

Pangasius

Cod

Crab

Catfish

Clam

Part of the problem is humans' fishing methods. One method, bottom trawling, involves dragging a large net across the ocean floor. This is meant to catch shrimp and fish but also picks up any other creatures in the way. The net causes damage to coral reefs and seagrasses too.

When young fish, mammals, turtles and sharks are caught accidentally, they are often thrown away. This can damage a whole generation of these creatures.

Unsustainable fishing is one of the greatest threats to marine life. Other dangers include pollution, habitat loss and global warming.

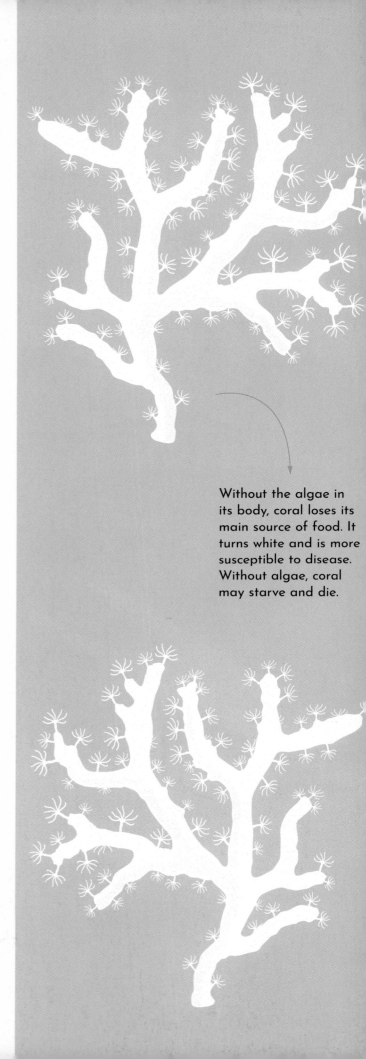

WARMING OCEAN

Earth's climate is changing, and the effects are still not totally clear. But from global warming to melting ice caps and rising sea levels, the impacts of climate change are already affecting the ocean.

Many human activities contribute to global warming. But changing these practices is difficult because they are so essential to how we live our lives. Finding a way to reduce them is one of today's big challenges.

Fossil fuels (coal, oil and gas) for energy or transport, farming animals and chopping down forests for wood all cause levels of greenhouse gases to rise. These gases, such as carbon dioxide (CO_2) and methane (CH_4), trap extra heat in the Earth's atmosphere. When the temperature around the world starts to rise, this is called global warming. The ocean absorbs almost all of this excess heat so waters start to warm up too. This puts marine life in danger, as temperatures that are too high can cause oxygen levels in the water to decrease.

WHITE CORALS

Coral is struggling, too. When stressed by warmer temperatures or pollution, coral expels the algae living inside it. The coral is now left without its most important source of food. The most obvious sign of this is whitening, known as coral bleaching. When coral suffers like this it can devastate whole ecosystems, since so many animals rely on it for food and shelter.

Rising carbon dioxide levels in the atmosphere are also changing the chemical composition of seawater. When the water absorbs the carbon dioxide gas, it becomes more acidic. Ocean acidification harms hard-bodied marine animals, including corals, crabs and lobsters. In these conditions, building tough shells and skeletons is more difficult for them.

When stressed, coral polyps expel the algae living in their bodies

Without the algae in its body, coral loses its main source of food. It turns white and is more susceptible to disease. Without algae, coral may starve and die.

★ **Did you know?**

Between 2014 and 2017, the Great Barrier Reef in Australia suffered the worst ever bleaching events in history due to warmer temperatures.

PLASTIC SOUP

Plastic is a man-made material used in everyday life. It is can be made into bottles, clothes, phones, cars, toys and much more. Plastic is everywhere.

But plastic has become a plague for the planet. In particular, the ocean is badly affected. Plastic pollutes beaches around the world. An estimated 7 million tonnes of waste flows into the ocean each year. That's the equivalent of dumping a rubbish truck-full of plastic into the ocean every minute.

Plastic waste that floats will travel long distances on ocean currents. In some locations, called gyres, these currents circle around in a loop, trapping the plastic.

Plastic trapped in this loop can accumulate into huge patches. The largest of these is the Great Pacific Garbage Patch. It covers an area of about 1.6 million sq km - that's three times the size of France. Scientists think that 80,000 tonnes of plastic waste have built up there.

Plastic can become lodged in the stomachs of animals, making them sick and sometimes killing them. Dolphins, seals, sharks and other large animals also become snared in old fishing nets and suffocate. Discarded fishing nets make up almost half of the waste in the Great Pacific Garbage Patch.

Sea turtles often mistake plastic bags for their favourite food, jellyfish.

Did you know?

In 2017, scientists discovered chunks of polystyrene (a type of plastic) lying on ice floes just 1,600 km from the North Pole. The area had previously been inaccessible to humans due to the permanent ice cover.

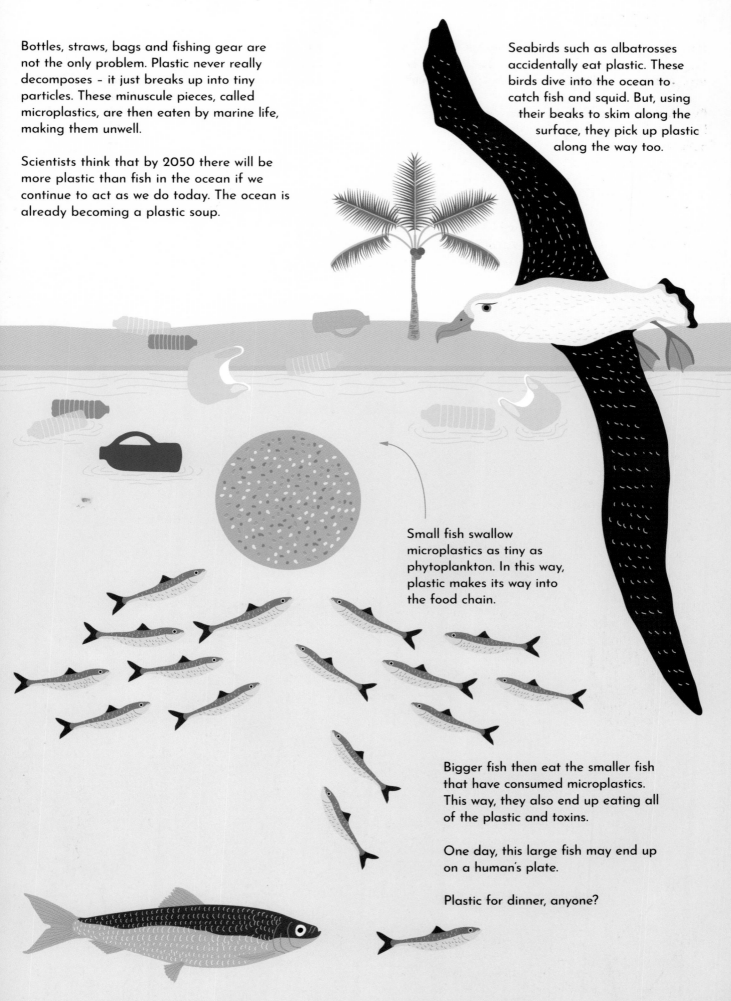

Bottles, straws, bags and fishing gear are not the only problem. Plastic never really decomposes - it just breaks up into tiny particles. These minuscule pieces, called microplastics, are then eaten by marine life, making them unwell.

Scientists think that by 2050 there will be more plastic than fish in the ocean if we continue to act as we do today. The ocean is already becoming a plastic soup.

Seabirds such as albatrosses accidentally eat plastic. These birds dive into the ocean to catch fish and squid. But, using their beaks to skim along the surface, they pick up plastic along the way too.

Small fish swallow microplastics as tiny as phytoplankton. In this way, plastic makes its way into the food chain.

Bigger fish then eat the smaller fish that have consumed microplastics. This way, they also end up eating all of the plastic and toxins.

One day, this large fish may end up on a human's plate.

Plastic for dinner, anyone?

How to Protect the Ocean

The ocean is in trouble – but it's not all bad news. There are lots of things you can do to make a difference!

SEE FOR YOURSELF

If you take a holiday, choose a trip which supports sustainable tourism. You can still go whale-watching or go scuba-diving on a coral reef, while caring for the environment!

LOOK, BUT DON'T TOUCH

When snorkelling or diving, never touch corals or other animals. They are fragile and easily damaged by human hands.

SPREAD THE WORD

Tell your family and friends what you've learned about the ocean. You might inspire them to protect the ocean more, too.

ADOPT AN ANIMAL

Many charities offer opportunities to sponsor threatened species such as whale sharks, polar bears and sea turtles.

BE A PICKY EATER

Next time you're at the supermarket, look closely at what you buy. The best seafood is local and sustainably caught.

SMALL IS BETTER

Eat less of big fish such as tuna, salmon and swordfish. These fish usually have long lifespans and it can take decades for their populations to recover from overfishing. Anchovies, sardines and other species further down the food chain are generally better options.

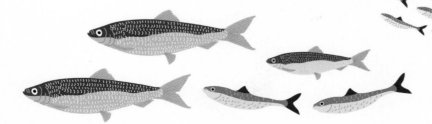

REUSE

Use only reusable bottles and bags.

VOLUNTEER

If you live near the ocean, volunteer in a beach clean-up.

SAY NO TO PLASTIC

At a restaurant or café, order a soft drink or juice without a straw.

Always try to avoid buying plastic bottles. When possible, ask for tap water or soft drinks in a glass.

RECYCLE

Sort out paper, glass, plastic and metal. They don't go in the normal bin.

PLANT A TREE

If you have a garden, planting a tree can help keep the air clean and fight climate change by removing carbon dioxide from the atmosphere.

KEEP IT HEALTHY

Reduce the amount of harmful chemicals that goes down the drains in your house. They will eventually make their way into the ocean.

Try to choose eco-friendly options for household items such as toothpaste, hand soap and body wash.

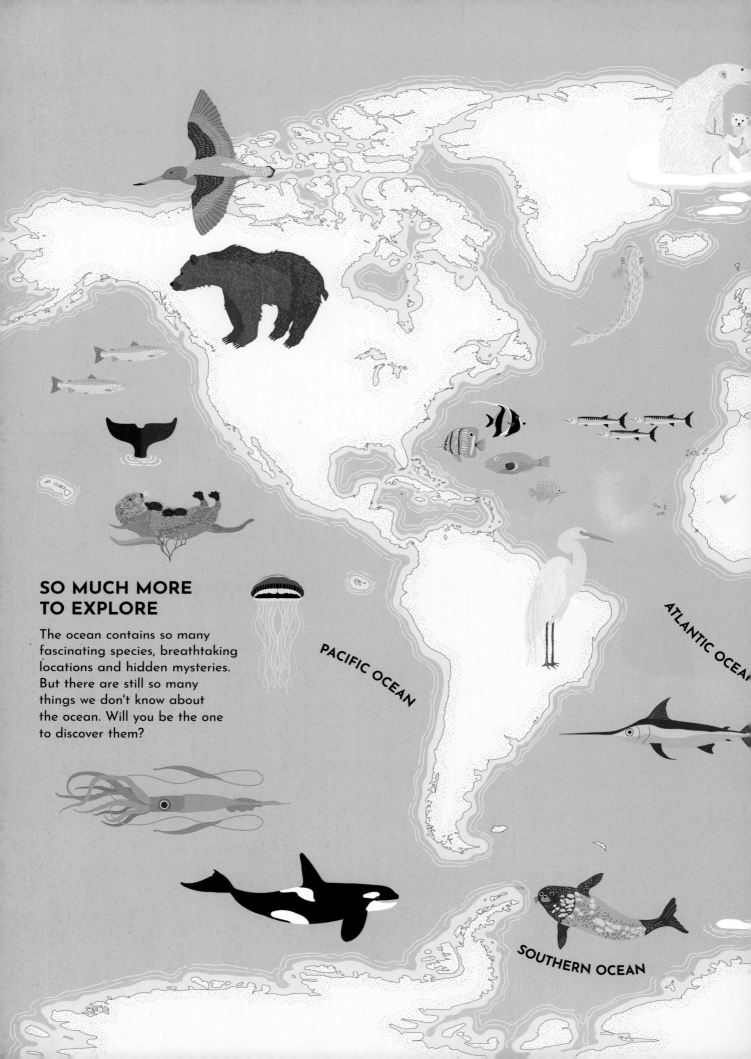

SO MUCH MORE TO EXPLORE

The ocean contains so many fascinating species, breathtaking locations and hidden mysteries. But there are still so many things we don't know about the ocean. Will you be the one to discover them?

PACIFIC OCEAN

ATLANTIC OCEAN

SOUTHERN OCEAN

ARCTIC OCEAN

INDIAN OCEAN

Glossary

A

ACIDIFICATION – increase in acidity level. In the oceans, acidity is increasing because more carbon dioxide (CO_2) gas is being absorbed from the atmosphere.

ALGAE – plantlike organisms that make their own food through photosynthesis.

ALPHA – animal in a group with the highest rank. Alphas dominate the other members of the group.

B

BIOFLUORESCENCE – ability of some marine organisms to absorb light and then to re-emit it in a different colour.

BIOLUMINESCENCE – production of light by some organisms such as fish, corals, plankton and bacteria. These organisms usually live in places that sunlight does not reach, such as the deepest parts of the ocean.

C

CARAPACE – top part of a turtle's shell. The carapce usually consists of bone plates covered in scales called scutes.

CLIMATE CHANGE – long-term alterations in weather patterns and conditions. Today, these are mainly caused by a rise of carbon dioxide (CO_2) and methane (CH_4) gas levels. These changes can affect average temperatures, precipitation and winds. As a result, ice sheets are sinking, sea levels are rising and extreme weather events are becoming more frequent.

CARBON DIOXIDE (CO_2) – gas that naturally makes up 0.04 per cent of Earth's atmosphere. But the amount of carbon dioxide in the atmosphere has increased since the Industrial Revolution, since humans began burning lots of fossil fuels (like coal and oil). Increasing carbon dioxide levels is one of the causes of climate change.

CRUST – outermost (and thinnest) layer of the solid rock covering the Earth. There are two distinct types of crust: oceanic crust and continental crust. Oceanic crust makes up the sea floor and oceanic trenches. On average, it is about 6 to 7 km thick. Continental crust mainly corresponds to land above sea level. It can be between 25 and 70 km thick.

CRUSTACEANS – group of invertebrates that includes crabs, lobsters and shrimps. They have jointed limbs and a hard outer shell.

E

EARTH'S MAGNETIC FIELD – natural phenomenon that makes Earth behave like a big magnet. Its two poles create invisible power lines. This magnetic field protects living creatures, including us, from some of the dangerous radiation from the Sun. Some animals, like salmon and turtles, can sense the magnetic field and use it to find their way.

ECHOLOCATION – ability of some animals, like dolphins, whales and bats, to detect prey and obstacles using echoes. The animals make a sound and then sense how long it takes to come back. They then use this information to work out how far away prey or obstacles are.

ECOSYSTEM – combination of living organisms and non-living elements interacting in a certain area.

ENDEMIC – organism that originates in a certain area.

EQUATOR – imaginary line running around the middle of the Earth at its widest point. The equator divides the globe into northern and southern hemispheres.

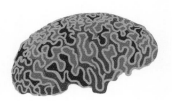

F

FOOD CHAIN – system that links organisms in an ecosystem by what they eat and what eats them. Nutrients are passed up the chain from one organism to the next.

G

GENES – information contained within DNA, a complex molecule found in the cells of all living things. Genes contain hereditary traits. This is the information that determines the features of an organism, such as its size, shape, colour and markings.

GILL – organ found in many marine animals, such as sharks and other fish. It allows them to capture the oxygen in the water that they need to survive.

GREENHOUSE GAS – any of several gases that are naturally present in the atmosphere, such as carbon dioxide and methane. Greenhouse gases trap the heat that the Earth emits through infrared radiation. The increase of these gases in the atmosphere due to human activity has made the temperature of Earth's atmosphere rise.

H

HABITAT – place or environmental conditions where an organism lives.

I

INVERTEBRATES – animals without an internal skeleton or backbone. Invertebrates include animals as diverse as crabs, sponges, sea urchins, snails, clams and squids.

M

MICROSCOPE – instrument that allows us to look at objects and organisms too small for the human eye to see.

MID-OCEAN RIDGES – mountain systems formed when two tectonic plates of the ocean's crust pull apart. Mid-ocean ridges are around 2,000 to 3,000 m tall.

MOLLUSCS – invertebrates with a soft body often protected by a shell. Mussels, octopuses, squids and sea slugs are marine molluscs. Snails and slugs are molluscs found on land.

MOUTH (RIVER) – place where a watercourse such as a river or stream flows into the sea or a lake.

O

OCEANIC TRENCHES – long, deep canyons in the sea floor. Oceanic trenches can be over 6,000 m deep, making them the deepest areas on Earth. They are formed when two plates of the oceanic crust crash together and one slides under the other.

ORGANISM – any individual living thing, such as an animal, plant or single-celled bacteria, that is capable of growth and reproduction.

P

PACK ICE – floating ice that originates from polar sea ice and drifts away from it.

PHOTOFORE – organ that can produce light. Photofores are usually found among species living in places where it is always dark.

PHOTOSYNTHESIS – process by which plants and some other organisms create their own food. They make a sugar called glucose from water, carbon dioxide and sunlight. They also produce oxygen in this process, which is then released into the air and is essential for most organisms' survival.

PLANKTON – group of organisms that drift with the ocean's currents. There are two kinds: phytoplankton

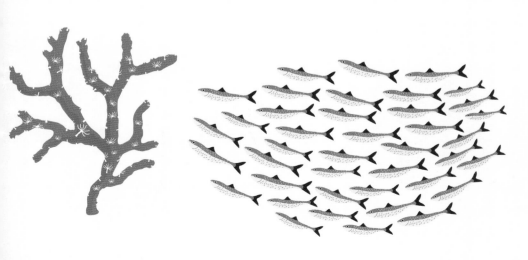

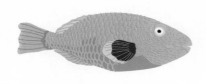

and zooplankton. Phytoplankton make energy through photosynthesis. They include microscopic algae and form the base of food chains in most aquatic ecosystems. Zooplankton are tiny animal organisms that feed on phytoplankton. They are very small and include both microscopic animals as well as the young of much larger fish and other marine creatures.

POD – group of marine mammals. Herd and school are other words for pod.

POLLUTION – contamination of the air, water or soil by man-made substances. These can include toxic chemicals, oil and plastics.

R

RESOLUTION – how precisely detail can be seen when looking at a picture or a map. The higher the resolution, the more details are visible.

S

SCAVENGERS – animals that feed on organisms that are already dead.

SCHOOL (OF FISH) – big group of fish from the same species that all swim together in a coordinated way.

SEDIMENT – rocks are broken down by oceans and rivers into smaller bits called sediment. This sediment is deposited on river beds or ocean floors, where it sits on top of the oceanic crust.

SONAR – tool that uses sound waves for navigation, communication, identifying objects (such as shipwrecks) and mapping the ocean floor.

SPECIES – basic group for classifying organisms. Individuals from the same species can reproduce and have fertile offspring.

SUBMERSIBLE – small vehicle used for underwater exploration.

SUSTAINABLE FISHING – method of fishing that protects the survival of marine organisms.

SYMBIOSIS – close connection between two or more organisms of the same or different species. If both organisms benefit from the relationship, it is called a mutualistic symbiosis. There are also types of symbioses where only one species benefits, or where the other may be harmed, as in the case of parasites.

T

TECTONIC PLATES – large, moving pieces of rock that make up Earth's outer layer. The chunks fit together like the pieces of a jigsaw puzzle. The movement of these plates can cause earthquakes and volcanic eruptions, while plates crashing together can form mountains.

V

VERTEBRATES – animals that have a skeleton and backbone. These include fish, reptiles, birds and mammals – including humans.

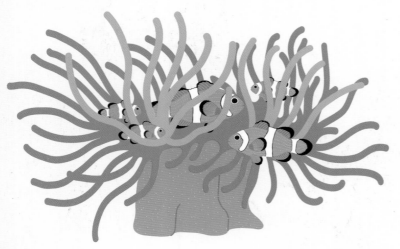

INDEX

What on Earth Books, The Black Barn, Wickhurst Farm, Tonbridge, Kent TN11 8PS, United Kingdom

© 2018 White Star s.r.l.
Piazzale Luigi Cadorna, 6 - 20123 Milan, Italy
www.whitestar.it

White Star Kids® is a registered trademark property of White Star s.r.l.

First published in English by What on Earth Books in 2019

Staff for this book:
This edition edited by Patrick Skipworth
Graphic design by Valentina Figus
Cover design by Andy Forshaw

A CIP catalogue record for this book is available from the British Library

ISBN: 978-1-9999680-5-2

Printed in Italy by Rotolito S.p.A. - Seggiano di Pioltello (MI)

2 4 6 8 10 9 7 5 3 1

whatonearthbooks.com

Sabrina Weiss

Originally from Switzerland, Sabrina is a London-based science communicator working for an environmental society and as a freelance writer. She previously worked in the non-profit, aviation and technology sectors, always tasked with presenting complex information in an engaging and understandable way.

Giulia De Amicis

Since completing her masters in communication design in 2012, Giulia has been working as a visual designer and illustrator. Her work mainly focuses on the display of information for newspapers, magazines and the environmental sector, with a particular interest in marine ecology, geography and human rights.

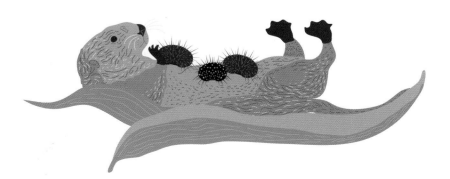

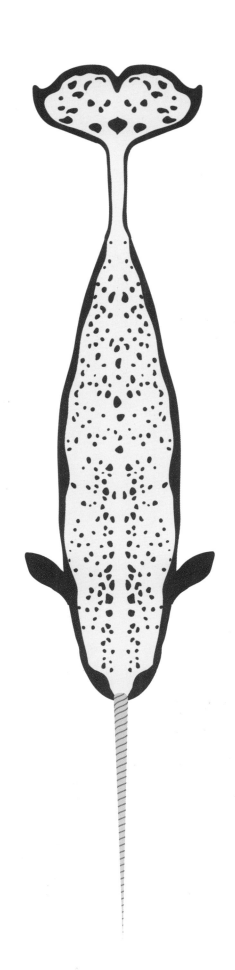